# NURTURING NATURE AND THE ENVIRONMENT WITH YOUNG CHILDREN

This book, at the intersection of early childhood and reconceptualizing practice, looks at how practitioners, theorists, and teachers are supporting young children to care about the environment differently.

Despite the current popularity of post-human perspectives, in social science more broadly and in early childhood studies more specifically, this is one of few to make visible international practices and perspectives that emerge at the intersection of early childhood education, environmental justice, sustainability, and intergenerational/interspecies communities. The book provides an innovative exploration of the links between children, elders, and nature. With contributions from established scholars, practitioners, and newcomers this book reframes educating for social justice within an ecological landscape; one in which young children and their elders are mobilized to understand, reconceptualize, and even undo negative environmental impact, whilst grappling with the ways in which the earthly forces are acting upon them. Specific theoretical chapters (spirituality, nature, critical and post-human/materiality, pragmatics, and constructivist approaches) are blended with applications of pedagogic strategies from across the globe.

This book responds to a growing interest among early childhood professionals and scholars for sustainably focused and ethically reimagined programs. This collection rewards the reader with opportunities to critically reflect on their own practice, delves into new terrestrial collectives, and explores new pedagogical pathways. It will be essential reading for practitioners and scholars alike.

**Janice Kroeger** is Graduate Coordinator and Associate Professor of Early Childhood and Teaching at Kent State University, USA.

**Casey Y. Myers** is Coordinator of Studio & Research Arts and Assistant Professor of Early Childhood Education at Kent State University, USA.

**Katy Morgan** is a Doctoral Candidate in Curriculum and Instruction at Kent State University in Kent, Ohio, USA.

# NURTURING NATURE AND THE ENVIRONMENT WITH YOUNG CHILDREN

## Children, Elders, Earth

*Edited by Janice Kroeger, Casey Y. Myers and Katy Morgan*

Routledge
Taylor & Francis Group

LONDON AND NEW YORK

First published 2019
by Routledge
4 Park Square, Milton Park, Abingdon, Oxon OX14 4RN

and by Routledge
605 Third Avenue, New York, NY 10017

*Routledge is an imprint of the Taylor & Francis Group, an informa business*

*British Library Cataloguing-in-Publication Data*
A catalogue record for this book is available from the British Library

*Library of Congress Cataloging-in-Publication Data*
A catalog record has been requested for this book

ISBN: 978-0-815-35924-1 (hbk)
ISBN: 978-0-815-35929-6 (pbk)
ISBN: 978-0-429-26467-2 (ebk)

Typeset in Bembo
by Swales & Willis, Exeter, Devon, UK

I Pledge Allegiance to the World,
To Care for Earth and Sea and Air.
To Cherish every Living Thing,
With Peace and Justice Everywhere.
Signed, *Earth Day Every Day*

**Lee Cheramy, Kindergarten Teacher, Leal School, Urbana Illinois**

# CONTENTS

# ILLUSTRATIONS

**Figures**

## Tables

# CONTRIBUTORS

**Janice Kroeger** is currently Associate Professor in Curriculum and Instruction and Graduate Program Coordinator for Early Childhood Education at Kent State University in Kent, Ohio, USA. Kroeger is a former child care worker, an inclusive educational specialist at Colonel Wolfe School in Champaign-Urbana, and an employee of Anita Purvis Nature Center in Urbana, Illinois. Kroeger has expertise in qualitative methodologies and early childhood education. Kroeger is best known for scholarship in family–teacher relationships and home–school community partnerships in public school settings with a focus on diversity and identity. Kroeger has published numerous articles about early childhood classroom practice, anti-bias curriculum, documentation, and social-emotional belonging in classroom communities, as well as cultural change related to political and social community action. Her research and teaching interests involve inclusive early childhood practices including sustainable futures. As a qualitative researcher, Kroeger uses an eclectic blend of conceptual frameworks including critical, discursive, material culture and identity focused theories to advance practice.

**Casey Y. Myers** is currently Assistant Professor of Early Childhood Education at Kent State University in Kent, Ohio, USA. She is also the Coordinator of Studio and Research Arts at the Kent State University Child Development Center, an early years laboratory school. Her research and teaching interests revolve around the everyday, more-than-human materialities of young children's school lives. She was the 2015 recipient of the Jeanette Rhedding-Jones Outstanding Dissertation Award from the Reconceptualizing Early Childhood Education organization.

**Katy Morgan** is a Doctoral Candidate in Curriculum and Instruction at Kent State University in Kent, Ohio, USA. Her focus in teaching and research is in Social Studies, including moral reasoning and historic empathy.

**Kate Albing** is an educator in the United States. Her teaching, research, and writing focuses on relational ontology and the connections between children and place. She is presently earning her M.Ed. in Early Childhood Education from Kent State University.

**Laila Aleksandersen Nutti** is an Associate Professor of Education at Sámi University of Applied Sciences. Her focus in teaching and research is on Yoik (Sámi traditional chanting), aesthetics, technology, and traditional knowledge within the field of Indigenous Early Childhood Education. She is the leader of the UArcitc Thematic network for indigenous teacher education.

**Solomon Amuzu** is the founder and managing director of Call to Nature Permaculture, a Ghana-based non-profit that uses permaculture ethics and principles to promote community development. Call to Nature Permaculture was founded to care for the Earth, care for people, and share resources by implementing permaculture principles and providing a high quality of permaculture education and training.

**Aslaug Andreassen Becher** is an Associate Professor of Education at Oslo Metropolitan University (Earlier: Oslo and Akershus University College), where her teaching and research focus on early childhood and primary education. Her research and publications are related to multicultural issues, including indigenous perspectives in early childhood institutions and teacher education, as well as children's spaces, bodies, and materiality.

**Anna Beckwith** is a graduate of Kent State University's Early Childhood Master of Arts in Teaching program and is pursuing a permanent teaching position.

**Terri Cardy** has taught courses in the early childhood education program at Kent State University and has been a lead preschool teacher at the Child Development Center laboratory school. Currently she is the Outdoor Educator at the Child Development Center. Her role is to collaborate with children and classroom teachers to support their work in the outdoors and to provide opportunities for children to strengthen their relationship with the natural world. She also has a strong interest in actively involving young children in community service work in the local community. Before coming to Kent State, she earned undergraduate and graduate degrees in Elementary Education and Early Childhood Education and taught in public schools and Head Start in Pennsylvania.

**Jacob Dunwiddie** is a graduate of Kent State University's Early Childhood Education Bachelors program and a graduate student at the University of Vermont studying Higher Education. Jacob's love for the outdoor education stems from his own outdoor experience in the primary years which were paved parking lots. At Kent State University's Child Development Center his love for outdoor education grew because of the opportunities offered to the children, staff, and faculty to engage in forest, wetlands, garden, and playground. Jacob hopes to return to the classroom and bridge the gap between the outdoors and education.

**Lynn Gregor** has over 25 years of experience in horticulture, specializing in urban agriculture and organic garden education. Lynn served as the OSU Extension Program Coordinator for ten years working alongside community gardeners in Cleveland, developing and maintaining gardens. There she gardened with juvenile offenders, established the market garden training program and co-founded City Fresh. She and her husband co-wrote the book, *A Place to Grow: Voices and Images of Urban Gardeners* (The Pilgrim Press, 1998). Currently, she is the Farm-to-School project consultant with the Akron Public Schools; spearheads her own Green Sprout Gardens, helping others to grow food for themselves and teaches at Old Trail School in Peninsula, Ohio.

**Karen Malone** is a Professor of Education and Research Director, in the Department of Education at Swinburne University of Technology in Melbourne, Australia. She is an international expert in early years environmental education focusing on childhood nature, multi-species relations, the posthuman subject, creative and participatory methodologies, and postqualitative research. She engaged with conceptual research theorizing through post-humanism and vital (new) materialism. Her latest research focuses on under two-year-olds and conceptualizes the notion of sensing ecologically.

**John T. Ng'asike** is a Senior Lecturer at Mount Kenya University in the department of early childhood studies. He has a Ph.D. in Curriculum and Instruction from Arizona State University in the USA. A Ford Foundation Fellow and a visiting scholar at Oslo University, Dr. Ng'asike has published widely in the field of early childhood education. His research interest is in the African concept of early childhood education.

**Nina Odegard** is a Ph.D. candidate at the Faculty of Education and International Studies at Oslo Metropolitan University, Department of Early Childhood Education. Her Ph.D. project is called "Aesthetic explorations of recycled materials – in light of materiality". She arrived at Oslo Met from a position as a project leader building a Creative Recycle Center for two municipalities and as a pedagogical supervisor (pedagogista) for the ECCs in Porsgrunn municipality, inspired from a Reggio Emilia approach. She has also written a book in Norwegian called *Reuse as a Creative Force: When Matter Comes to Matter*. Before being a project leader she worked for 16 years in ECCs, mostly as a center manager.

**Adonia Porto,** Ph.D. is a lead preschool teacher at the Kent State University Child Development Center, an early years laboratory school and a lecturer in Early Childhood Education. Her research and practice interests revolve around children's experiences with/in nature.

**Lolagul Raimbekova** was born and raised in a mountainous region of Tajikistan, known as Gorno-Badakhshan Autonomous Oblast (GBAO). She completed her Diploma of Specialist in Teaching English as a Second Language in her hometown Khorog and taught English as a Second Language for college students for more than eight years. She is currently pursuing her Ph.D. at Kent State University, Kent Ohio, USA in Curriculum and Instruction. Her major research interest includes globalization, family diversity, as well as supporting pre-service and in-service teachers in their work with international children.

**Abigail E. Recker**, M.Ed., is a pre-K–grade 3 teacher in Orono, Maine, USA. Abigail's research focuses on fostering a culturally sustaining community of learners through authentic inquiry in elementary settings.

**Jenny Ritchie** has been involved in the early childhood care and education sector since the 1970s, as a childcare worker, kindergarten teacher, parent, teacher educator, education researcher, and more recently, a grandparent. She is an Associate Professor in Te Puna Akopai, the School of Education, at Te Whare Wānanga o te Ūpoko o te Ika a Māui, Victoria University of Wellington, New Zealand. Her research and teaching has focused on understanding how to apply a commitment to Te Tiriti o Waitangi within early childhood and teacher education and exploring ways in which applying Māori conceptualizations can enhance pedagogies that support and care for our planet.

**Aubrey Ryan** is a graduate of Kent State University's Early Childhood Education Bachelors program and is pursuing a permanent teaching position.

**Jonathan Shaw** is a Kindergarten teacher in Northeast Ohio, USA. He received a Bachelors and Masters in Early Childhood Education from Kent State University. Jonathan focuses on the classroom environment and design, curtailing to students needs through developing classroom spaces and building child centered furniture. Currently Jonathan is working on developing outdoor natural learning spaces for his students and has been teaching wilderness survival skills to youth in Northeast Ohio for 15 years.

**Bushra Fatima Syed** is an assistant Professor at Oslo Metropolitan University. She teaches at Faculty of Education and International Studies, Department of Early Childhood Education. Her current research interests include Norwegian kindergarten teachers' understanding and pedagogical approaches to sustainable development. Her perspectives are emerging from multiculturalism, critical multiculturalism theory and inspired by postcolonial theories.

# INTRODUCTION

## Why nurture nature and the environment with young children?

*Janice Kroeger, with Casey Y. Myers*

The central foci of this book, which traverse the boundaries of places, research approaches, and pedagogies, are conceptual choices in teaching and learning that enhance children's understanding and centrality of the Earth within the daily life of the early years, schooling, and curriculum. In this book, *Nurturing Nature and the Environment with Young Children*, we take a philosophically grounded approach that merges multiple theoretical perspectives such as critical, indigenous, feminist, post-colonial, and other post foundational ideologies and methods. The practices in this book occur in a variety of locales, as researchers and practitioners around the world elevate the care and preservation of species, clean air and water, stewardship of and support for natural and sustainable resources, as well as understandings of the agency of non-human beings within the lives of young children.

The Earth we share is important; more important, in fact, than any *one* person. Thus, in this book we think of children in relation to Earth, elders in relation to Earth and children, and Earth in relation to non-humans and so on. In a Gregorian knot-like configuration the intractable problems and possible solutions of the present reside. The collection's subtitle resounds because *Children, Elders, Earth* isn't a book of "how to" approaches from practitioners and researchers, but rather a collection of *witnessing*, showing those inspired, unusual, often catastrophic futures, and also those *better-imagined* possible sustainable futures that some of us hope for our children *now*. A central premise of the text is the use of intergenerational, community-based practice to further children's care of their environments and multi-species companions, as well as a concern for living matter(s) upon which all other life depends.

## Learning how to sweat

In this introduction chapter, I (Janice) use *learning to sweat* as a metaphor for the intellectual dissonance and often bodily discomfort I have experienced as I (we) contemplate the politics of responsibility of teaching young children and those who care for them in *this* world, a world characterized by an absence of value to the soul of *wakan*, non-speaking living species and colonization of first people(s) (Latour, 1993; Shotwell, 2016). Today as in other times, in North America the *Sweat* is an ancient healing ceremony among Sioux, Lakota, Cree (and many other North American Indians) who are now mostly gone or have relocated over centuries of colonialist occupations. Geographically, the editors of this book reside in North America, where the Osage, Potawatomi, Cherokee, Shawnee, and Seneca were some of the first people(s). Native Americans believe in *Wakan*, an articulation of the holy men about the bound-togetherness of life forms or "all that moves" (Walker, 1991, p. 81); first peoples in North America were the first to argue that human activity has created tremendous burdens upon the Earth, sickening it (Tyon, Grey Goose in Walker, 1991).

I liken the experiences felt in the sweat lodge to the transformations and challenges that are upon us as educators around the globe who do early childhood curriculum. Although the ethics and politics of worldly responsibility are not new to indigenous groups (Ritchie, Becher et al. this edition), we turn to philosophers Alexis Shotwell (2016) and Bruno Latour (1993, 2017) as well as others in this collection to weave together a formulation of thoughts to conceptualize our collection within childhood. Borrowing from the Lakota (a first people), *Wakan* is a centering of the air, water, breath, animals, plants, and other living matter as *sacred*, and this, although unorthodox in an academic text, helps us to re-center our thinking about how our healthy interconnections remain vital and essential to the well-being of *all* (Walker, 1991).

Indigenous knowledge systems—often important carriers of values for children about the interconnectedness of the world, the child, and the community—have heretofore gone missing as globalization and colonialism have dominated, leaving a gap of knowledge for the young to understand how their actions might impact adversely *or protect and care for* the Earth. Contemporary philosopher of science Bruno Latour argues (1993, 2018) that the emergence of science and the emergence of modernity has created irrevocable divides between nature, culture, and human knowing (2018, p. 11). One result of the rise of science (the moderns) was the rush to divide it irrevocably with the past (the pre-moderns). Thus, the balance of ecosystems, species, and humans within the world has been disrupted. While we recognize communities that have long reclaimed and centralized their original knowledge(s) within early childhood education and value them greatly (Rau & Ritchie, 2014; Rinehart, 2005; Taylor, 2013, 2018, we also acknowledge that our indigenous elders in many parts of the world are gone, and their knowledge(s) lost.

I use the metaphor of "the sweat" to justify the task of curriculum as a more-than-human endeavor, not because I have heretofore considered myself a post-humanist, but instead because I am sweating as a human in a world that might not make it, in what Haraway has called the "Chthulucene." Additionally, as a human who has recently taken up the sweat lodge as a spiritual practice to cope with my own humanity, the learning I have experienced provides a sometimes humorous, sometimes painful, and more than uncomfortable set of descriptions of the body undergoing the sweat; for my teachers have "jogged" for me a primordial memory (the ooze of water as it hits searing hot stones), that prays to our relations—the trees, the water, the stones (an embodiment of the grandfathers), the oldest things on the planet—to take pity on us as we as a species undergo transformation in consciousness and in behavior.

## Reconsidering from whom children learn: elders and other(s)

Our consideration of "elder" is broad in this collection, as we try to reposition the Earth (and care for it) within our focus on teaching and learning. As a vehicle to this worldly inclusion, the position of "elders" includes the multiple-aged, as apprenticing and practiced "experts" who co-mingle with children. We allow viable models to be shared with readers who are interested in doing similar types of work in their classrooms, schools, families, and communities. Elders may include community resources from beyond the school or classroom with potential expertise(s) in project areas: master gardeners, tribal elders, clan leaders, matriarchal figures, farmers, parents, laborers, grandparents, farmers' market vendors, lay people, and the scrapper(s). Teachers/Elders might be conceptualized as matter or living beings. Elders might also include older students or children themselves as they mentor each other. Ultimately, we acknowledge that social change of impact is often constructed within communities when individuals from very different, cross-cutting groups are working together within a broad construction of community, and finally moving toward collectives that include more than humans (such as the dirt, the compost, the worms, our shelters, and our adjacent meadows for which we care) (Kroeger, 2006; Lash & Kroeger, 2018; Kroeger, Cardy, Recker, Gregor, Dunwiddie, & Beckwith, this edition). In this book, we juxtapose teachers, scholars, "elders" from around the globe.

The reasons for the collection of work in this book are many, but we as editors and chapter contributors are primarily answering a need in the field of early childhood education for pedagogies and practices as well as frames of thinking that *de-center* our largely humanist framework of social science and instead "include the earth, nature, the entire human and non-human milieu, as equally important" as we go about our daily lives in universities and classrooms (Pacini-Ketchabaw & Nxumalo, 2018; Taylor, 2013; Tesar & Arndt, 2018, p. 116). We note the post-human moves in reconceptualizing early childhood circles in Hong Kong (Taylor, Blaise, Giugni, 2012), then in London (Bone, 2010) and again in Kenya (Becher, this edition). Particularly, for me, this work has evolved as many of the (now former) students I have worked with have been especially interested in materiality and research arts with young children, children's perception and experience of natural spaces, spirituality, and eco-pedagogic capacities. In the last several decades, environmental educators have articulated the importance of the early childhood sector as a niche for considering the timeliness of young children's understanding of the environment and nature (Davis, 2008; Hedefalk, Almqvist & Östman, 2014). Approaches for new thinking in early childhood education and sustainability reveal "a more competent child that can think for themselves and make well considered choices" (Hedefalk et al., 2014).

The present time is what many in circles of posthuman thinking have called the Anthropocene (Haraway, 2016; Malone, 2018). Human activities—particularly our use of fossil fuels and the proliferation of late capitalism—have transformed our planet in harmful, irreparable ways (Haraway, 2016; Latour, 1993, 2018; Shotwell, 2016, pp. 2–3). The Anthropocene (according to Haraway) is a time of

> multispecies, including human urgency: of great mass death and extinction; of onrushing disasters, wholly unpredictable specificities ... of refusing to know and cultivate the capacity of response-ability; of refusing to be present in and to onrushing catastrophe in time; *and at the same time* (our addition) of unprecedented looking away.
>
> *(2016, p. 35)*

Anthropocene for early educators, then, is both a real *now* and an imagined future reality; Haraway calls us to be "fully" present "with the trouble." Haraway reminds us,

> In urgent times, many of us are tempted to address problems in terms of making an imagined future safe, of stopping something from happening that looms in the future, of clearing away the present and the past in order to make futures for coming generations.
>
> *(2016, p. 1)*

But she argues that we don't have to do that; "staying with the trouble" means learning to be truly present.

Today, global crisis, climate instability and change, population density, and tremendous challenges to the air, the water, and the land are upon us, causing us to necessarily rethink and retool our educational priorities with young children (Boix-Mansilla, 2011; Kahn, 2010; Malone, 2018; Prince, 2010). Recently, for example, the United Nations has called upon the world to respond with urgency to matters of climate disaster, giving policy makers and nations approximately 11 years to change the trajectory of policy landscapes in order to mitigate human catastrophe. Intercultural educator Boix-Mansilla has long argued that one of the most important skillsets children in this century will need to learn is *climate stewardship* in a world of climate instability. That is, "students today will need to understand the earth and how it works as well as their impact upon ecosystems and humanities' dependence upon the earth for life" (Boix-Mansilla, 2011, p. 20). But *what* might this mean for early educators and those who work with young children? Duhn (2012) argues that our most pressing issue is "learning to care for all life on earth, when topics like climate change" lodge early childhood educators and young children *out of innocence* as we "enable children to participate and contribute to the issues that affect their lives now and in the future" (p. 3).

We have entered, and will lead, a new era of social justice pedagogies in early childhood (Ramsey, 2014; Ritchie, 2013). Stepping back and being present in our work with children and Earth means knowing where we've been and where we can go next as a field and as teacher educators. Challenging mainstream teaching pedagogies that focus upon the individual by way of scores, testing, and curriculum as mandated has become central to reforming teaching and learning (Taylor, 2018). More fully, however, early childhood educators around the globe are responding to and reassessing the "potential risks and consequences to operat[ing] within an unreflexively human-centric and overtly individualistic pedagogical framework" (Taylor, 2018, p. 212). Collectively in this text we are responding to Taylor's (2013) commentary that

> twenty-first century children need relational and collective dispositions, not individualistic ones to equip them to live well within the kind of world that they have inherited … Such dispositions and capacities will never be fostered through the application of a child-centered and hyper-individualistic developmental framework, nature-loving or not.
>
> *(p. 117)*

Our purpose in this book, then, is to provide a challenge and transformation to everyday educators of young children and teacher leaders. Duhn (2012, p. 21) articulates clearly what task is at hand in early childhood education:

> Fostering ethics of care requires that adults challenge dominant constructs of childhood by enabling children to participate and contribute to the issues that affect their lives now, and in the future. Early childhood teachers who take the call for engagement seriously take on a leadership role (Duhn, 2010) …. If children are humanity's hope for the future (Chawla, 2001), reconceptualizations of childhood are a pressing issue. For early childhood education, this means that rethinking childhood has to be a core aspect of eco-focused pedagogies.

We have created this collection in solidarity with others, offering what we hope is a selection of well-chosen chapters that challenge what pedagogy and research can look like, while acknowledging that there are countless others around the globe who are doing—and will be inspired to do—the same.

## Changing conceptualizations of justice: from social to eco justice

Reaching beyond the reconceptualist realities of the present day, in which inclusivity has often been a local endeavor to support and re-imagine individuals and communities from marginalized subject positions of race, ethnicity, economic status, gender and sexuality, language, and other intersecting identity statuses (Bloch, Swadener & Cannella, 2018; Genishi & Goodwin, 2008; Kroeger, 2006), this book acknowledges the critical indigenous, feminist, and post-colonial and *now post-human* understandings *necessary* to address curriculum with young children in light of planetary crisis and education for social justice within an *ecological* landscape (that includes both the young child and others). Reconceptualist early childhood researchers have heretofore privileged identity and agency as local endeavors to support ethnically, linguistically, and gender-inclusive views of activist education, often calling upon those in the field to move beyond dominant pedagogical constraints to provide and re-imagine curriculum and policy that re-positions marginalized *individuals* and *communities* (Bloch, Swadener & Cannella 2014; Genishi & Goodwin, 2008; Tobin, Arzubiaga & Keys Adair, 2013). Furthermore, inclusivity is often aimed at helping children "belong" by providing frames of gender, sexuality, class, tribal clan, religion, or cultural background often at the privileging of "the exclusively human" concepts. Such works have furthered our thinking—but not enough as we consider the needs of the Earth. Our conceptual framing of care is to be corrected to form a broader "understanding of justice," including the Earth (p. 118, Taylor in Bloch et al., 2014, Ramsey, 2014).

In this collection, we have included the world beyond the child in substantive ways, arguing for pedagogical choices that educate, allow, and center the Earth, its creatures, and its needs in congruence with that of the human, in ways that account for the tremendous challenges facing Earth and her humanity. We argue in this edited volume for moving fully toward peaceful coexistences both within human and non-human realms, but centrally locate the Earth, its animals, plants, local and global ecosystems, and spiritual arrangements, examining how those peaceful priorities help us best educate and care for our youngest. What might it mean for each of us as we use intergenerational, community-based practices as a means of furthering young children's care for the environment, species, and engagement(s) with living and nonliving matter(s)? We recognize communities that have long reclaimed and centralized indigenous knowledge(s) within early childhood and value them greatly (Rau & Ritchie, 2018; Rinehart, 2005; Taylor, 2013, 2018) but we see in many parts of the world that our indigenous elders are gone, their knowledge(s) erased; as *we* struggle to reclaim and catch up, to be more humane and "other worldly" in our ethics of care, especially as those in North America, we ask how our curriculum action should "look." Given our awareness of planetary crisis, escalating climate change and the agency of the Earth, along with migration, displacement, and overwhelming negative human impacts upon the globe, how can we as adults (in some cases elders and children) be acting, doing, and advocating with young children?

## Conclusions: thinking in urgencies versus emergencies

Justification for such an endeavor like this collection accounts for the tremendous shift that has occurred as global educators include the current "worldly" challenges of unprecedented migrations, often overwhelming negative human impact(s) upon natural resources, and escalating global conflict(s). Justification for our collection acknowledges also that Earth damage is far reaching. Some geographic and social locations for children are affected more unfairly and more often by contaminated or destroyed resources. Conflict over ecological resources such as healthy food, clean water, habitable spaces continue, and conflicts regarding which humans go where on our shared Earth persist. In a time that many of us wish were characterized by interdependence rather than exploitation, this collection aims to help children develop new skill sets for problem solving and "getting along with others" as we ramp up the stewardship of the planet (Boix-Mansilla & Jackson, 2011; Taylor, 2013, 2018).

In her work, Haraway looks for stories in which multi-species endeavors become players who create *partial recuperations* of getting on together in just ways. Therefore, in this collection, we consider what it would look like if teachers cared for the Earth in the same way they have traditionally cared for young children. We also consider what curriculum would allow if children themselves were taught to care for the Earth and others (non-human) in the same way we prioritize care among peers. In our collection, we begin to answer how the trees, the rocks, the grass, living matter of dirt, or non-living matter of trash may be cared for in new collectives of action. Additionally, we conjecture how elders could learn from children and support the ways that children learn best as each of us cares for an Earth that commands central attention in educational work.

## References

Bloch, M. N., Swadener, B. B., & Cannella, G. S. (2014). *Reconceptualizing early childhood care and education—A reader: Critical questions, new imaginaries and social activism.* New York: Peter Lang.

Bloch, M. N., Swadener, B. B., & Cannella, G. S. (Eds.) (2018). *Reconceptualizing early childhood care and education—a reader: Critical questions, new imaginaries and social activism.* New York: Peter Lang.

Boix-Mansilla, V., & Jackson, A. (2011). *Educating for global competence: Preparing our youth to engage in the world.* Report available at www.asiasociety.org/education

Bone, J. (2010). Play and metamorphosis: Spirituality in early childhood settings. *Contemporary Issues in Early Childhood, 12*(4), 420–417.

Chawla, L. (2001). *Growing up in an urbanizing world.* London: Earthscan.

Davis, J. (2008). Revealing the research "hold" of early childhood education for sustainability: A preliminary survey of the literature. *Environmental Educational Research, 15*(2), 227–241.

Duhn, I. (2010). Professionalism/s. In J. L. Miller & C. Cable (Eds.), *Professionalization, leadership and management in the early years* (pp. 133–146). London: Sage.

Duhn, I. (2012). Making "place" for ecological sustainability in early childhood education. *Environmental Education Research, 18*(1), 19–29.

Genishi, C., & Goodwin, A. L. (2008). *Diversities in early childhood education: Rethinking and doing.* New York: Routledge Falmer.

Haraway, D. J. (2016). *Staying with the trouble: Making kin in the Chthulucene.* Durham, NC: Duke University Press.

Hedefalk, M., Almqvist, J., & Östman, L. (2014). Education for sustainable development in early childhood education: A review of the research literature. *Environmental Education Research, 21*(7), 975–990.

Kahn, R. (2010). *Critical pedagogy, ecoliteracy, and planetry crisis: The ecopedagogy movement.* New York: Peter Lang.

Kroeger, J. (2006). Stretching performances in education: Gay activism and parenting impacts identity and school change. *The Journal of Educational Change, 7*, 319–337.

Lash, M., & Kroeger, J. (2018). Seeking justice through social action projects: Preparing teachers to be social actors in local and global problems. *Policy Futures in Education, 16*(6), 691–708.

Latour, B. (1993). *We have never been modern.* Cambridge, MA. Harvard University Press. Trans. Catherine Porter.

Latour, B. (2017). *Facing Gaia: Eight lectures on the new climatic regime.* Medford. MA: Polity Press.

Latour, B. (2018). *Down to earth: Politics in the new climatic regime.* Cambridge, UK. Polity Press.

Malone, K. (2018). *Children in the Anthropocene: Rethinking sustainability and child friendliness in cities.* London: Palgrave Macmillan.

Pacini-Ketchabaw, V., & Nxumalo, F. (2018). Posthumanist imaginaries for decolonizing early childhood praxis. In M. N. Bloch, B. B. Swadener, & G. S. Cannella (Eds.), *Reconceptualizing early childhood education and care—A Reader: Critical Questions, New Imaginaries and Social Activism* (pp. 215–225). New York: Peer Lang.

Prince, C. (2010). Sowing the seeds: Education for sustainability within the early years curriculum. *European Early Childhood Education Research Journal, 18*(3), 423–434.

Ramsey, P.G. (2014). *Teaching and learning in a diverse world: Multicultural education for young children* (4th ed.). New York: Columbia University, Teachers College.

Rau, C., & Ritchie, J. (2014). *Ki te Whai au, kit e ao Marama*: Early childhood understandings in pursuit of social, cultural, and ecological justice. In M. N. Bloch, B. B. Swadener, & G. S. Cannella (Eds.), *Reconceptualizing early childhood education and care—a reader: Critical questions, new imaginaries and social activism* (pp. 109–120). New York: Peter Lang.

Rau, C., & Ritchie, J. (2018). Ki te Whai au, ki te ao Marama: Early childhood understandings in pursuit of social, cultural, and ecological justice. In M. N. Bloch, B. B. Swadener, & G. S. Cannella (Eds.), *Reconceptualizing early childhood education and care—a reader: Critical questions, new imaginaries and social activism* (2nd ed., pp. 195–204). New York: Peter Lang.

Rinehart, N. M. (2005). Native American perspectives: Connected to one another and to the greater universe. In L. Diaz Soto (Ed.), *The politics of early childhood education* (pp. 135–143). New York: Peter Lang.

Ritchie, J. (2013). Indigenous onto-epistemologies and pedagogies of care and affect in Aotearoa. *Global Studies of Childhood, 3*(4), 395–406.

Ritchie, J. (2016). Qualities for early childhood care and education in an age of increasing superdiversity and decreasing biodiversity. *Contemporary Issues in Early Childhood, 17*(1), 78–91.

Shotwell, A. (2016). *Against purity: Living ethically in compromised times.* Minneapolis, MN: University of Minnesota Press.

Taylor, A., Blaise, M., & Giugni, M. (2012). Haraway's "bag lady story-telling": Relocating childhood and learning within a "post-human landscape." *Discourses: Studies in the Cultural Politics of Education, 34*(1), 48–62.

Taylor, A. (2013). *Reconfiguring the natures of childhood.* New York: Routledge.

Taylor, A. (2018). Situated and entangled childhoods. In M. N. Bloch, B. B. Swadener, & G. S. Cannella (Eds.), *Reconceptualizing early childhood education and care—a reader: Critical questions, new imaginaries and social activism* (2nd ed., pp. 205–214). New York: Peter Lang.

Tesar, M., & Arndt, S. (2018). Posthuman childhoods: Questions concerning "quality". In M. N. Bloch, B. B. Swadener, & G. S. Cannella (Eds.), *Reconceptualizing early childhood education and care—A reader: Critical questions, new imaginaries and social activism* (2nd ed., pp. 113–128). New York: Peter Lang.

Tobin, J., Arzubiaga, A. E., & Keys Adair, J. (2013). *Children crossing borders: Immigrant parent and teacher perspectives on preschool.* New York: Sage.

Walker, J. (1991/1980). *Lakota belief and ritual.* (Eds. R. J. DeMallie & E. A. Jahner). University of Nebraska Press.

# PART I
# Worldly longing(s)

# 1

# UNEASY ASSEMBLAGES OF CHILDEARTHBODIES

*Karen Malone*

## Uneasiness

Precarity flourishes as the uncertainty and unpredictability of the current state of the planet continues to be the most pressing issue of this generation. The impact of climate change, habitat destruction, overpopulation, radiation, and human consumption means the sixth mass extinction in Earth's history is under way and it is thought to be more severe than previously feared. Over 50 years ago, Rachel Carson (1962) warned humanity dangerous chemicals and radioactive particles were causing increasing irreversible harm to all living beings.

> Only within the moment of time represented by the present century has one species — man— acquired significant power to alter the nature of his world. During the past quarter century this power has not only increased to one of disturbing magnitude but it has changed in character. The most alarming of all man's assaults upon the environment is the contamination of air, earth, rivers, and sea with dangerous and even lethal materials.
>
> *(p. 3)*

The impact of environmental pollutants on children's bodies, especially in large major cities, causes young children, nonhuman animals, and plants to die in increasing numbers. Due to their immature cells and closeness to the source (the earth) the build-up of toxins in their bodies is ingested at rates exponentially higher than adult humans.

This chapter takes you on journey to visit children in Semipalatinsk, Kazakhstan; a city where their everyday lives collide with stories embedded in dusty lively streets and playgrounds with historical traces of radiation. They bring to our attention a world that is in the making, and had been in the making since the atom bomb was first detonated as a form of controlling and managing humans and nature. The stories highlight the vulnerability of children in cities at times of planetary nuclear disasters as expressed through a concept of porosity as a significant material form that through diffractive theorizing has the potential to be an assemblage of a reconstituted ecological entanglement. It speaks deeply of a past, existing, and future life in an Anthropocentric world.

## Unsettling

The Anthropocene was a term coined first by Nobel Prize-winning atmospheric chemist Paul Crutzen and biologist Eugene F. Stoermer to describe the significant and irreversible impacts of human activities on Earth and the atmosphere described by Carson and others many decades before (Crutzen & Steffen, 2003). The Anthropocene as a rupturing force brings our attention to humans who are neither exempt from the ecological world nor exceptional to those we are acting/being/dying in relation with. Exploring the Anthropocene story is to speak of how humans became such a potent environmental force that a signature of all our doings, for good or ill, are measurable in the layered rock for millions of years to come. By altering climate, landscapes, and seascapes, as well as flows of species, genes, energy, and materials, we have damaged our planet, many say beyond redemption.

Many scientists proposed dates for when this new epoch would begin, one accepted start being in the 1950s, when human activity, namely rapid industrialization and nuclear activity, set global systems on a different trajectory. Scientists say nuclear bomb testing, industrial agriculture (particularly carcinogenic chemicals), human-caused global warming, and the proliferation of plastic waste across the globe have so profoundly and deliberately altered the planet from its natural state it should be marked by the renaming of this epoch. As Carson wrote in 1962, the changing planet was up until the past two centuries of human intervention, a series of natural events. Life including the earth's animals and plants were until this time molded by the earth, an interactive dance of survival and adaption led by planetary evolution. Chemicals and other lethal materials produced by modern society have set off a chain of evils where life now affects the planet, irreversible and universal contamination seeping into all aspects of living tissues—radiation being one of these central lively actants.

> Radiation is no longer merely the background radiation of rocks, the bombardment of cosmic rays, the ultraviolet of the sun that have existed before there was any life on earth; radiation is now the unnatural creation of man's tampering with the atom.
>
> *(Carson, 1962, p.3)*

As an unsettling ontology, the notion of the Anthropocene disrupts a persistent "humanist" paradigm in disciplines such as education by allowing new conversations to emerge around human-dominated global change, human exceptionalism, and the nature/culture divide (Lloro-Bidart, 2015). As a disrupting ontological tool it reveals there is no homogenous/universal species and the scale and impact of ecological damage is unequal, unethical, and unjust; indigenous peoples, woman, children, and the other-than-human species we share this planet with are in it more than those entrenched in dominant western masculine cultures. Were we asleep at the wheel while corporations metastasized into these monstrous creatures of capitalism? Did we ignore the clarion call of the Anthropocene? There has been critique from many in regard to the naming of the Anthropocene. One argument has been its universalist nature. Universalism produces an assumption that we (humans/nonhumans) are all in this together and implicated in a balanced and uniform manner. This universalizing of the human predicament neglects to acknowledge the extent of diversity in the human/nonhuman experience and the ways in which wealth, nationality, ethnicity, gender, class, age, location and so on mediate relationships with the planet (Malone, 2018). And that the burden of the Anthropocene overpopulation, limits to growth, is often placed at the feet of the most impoverished even though they often contribute the least to its manifestation. That is, the scale of human ecological impact is unequal, unethical, and unjust; the poor, the children, and the nonhuman are more in it than the wealthy (Malone, 2018).

The uptake of radioactivity associated with the proliferation of nuclear weapons testing in the mid-20th century, for example, has been identified as one of the golden spikes indicative of the era of the Anthropocene. The nuclear age has left an invisible but global and affective reading of radiation, by employing the disruptive concept of porosity as a means for revealing our shared fragility—exposing our naked bodies and providing for an undressing of the exceptionalism of humans. Radiation is the effect of an entangled mattering of materials, objects, and bodies.

> Strontium 90, released through nuclear explosions into the air, comes to earth in rain or drifts down as fallout, lodges in soil, enters into the grass or corn or wheat grown there, and in time takes up its abode in the bones of a human being, there to remain until his death.
>
> *(Carson, 1962, p.3)*

Massumi (2015), drawing on the work of Spinoza, speaks of the "body in terms of its capacity for affecting or being affected" (p. 3); "to affect and be affected is to be open to the world, to be active in it and be patient for its return activity" (p. ix). At this time of the Anthropocene we are "in a far-from-equilibrium situation" (Massumi, 2015, p.114); we are beings affected and affecting the complexity of our times—this attunement to the "experience of precarity" brings with it chaotic situations, uneasiness, uncertainty. Those systems, our "bodies" (in its broadest sense) we have relied on, are in catastrophe and "there's no vantage point from which to understand it from the outside. We are immersed in it" (Massumi, 2015, p. 114). We are it, it is us. It is in us and we are in it.

## Diffraction

Exploring the complexities of children's lives in the Anthropocene, attuning to their entanglement within an assemblage of human–nonhuman matter, this is the work I am doing. I am queering awkward binaries—human/nature, subject/object, I self/other not self, adult/child—through diffractive theorizing by working with Barad (2007, 2014), Nancy (1991), Derrida (2005), and Smith (2013) as the means for interrupting discourses of human exceptionalism. Posthumanist approaches have the direct task of de-centering the human; it problematizes the notion of human as exceptional. The exceptional human assumes what matters to humans is the most important, and what matters to other species and things matters less. Posthumanist approaches demand a disruption of the human story, that we are somehow exempt from the consequences of our own contaminating ways— such an approach demands an "unlearning" of anthropomorphic ways of being and knowing the world, an onto-epistemological recasting of difference, a queering of binaries through diffractive theorizing. The focus of my recent research work is the onto-epistemological study of "lively matter—radiation" in the cities of Semipalatinsk in Kazakhstan (see Malone, 2018 for more specific details on the purpose of this project). I am curious to consider how radiation is entangled with humans and the collective of human–nonhuman things that are tied together; knotted in knots in an intricate ecological collective. Onto-epistemology assumes epistemology and ontology are mutually implicated "because we are of the world," not standing outside of it. I am working with Donna Haraway's (2003) notion of relational natures of difference and Karen Barad's (2007, 2014) tools of diffraction—not to map where differences appear, but rather map the effects of difference. I explore the technique of diffraction as an analytical tool by exposing the paradoxical potential of the diffraction of radioactive waves that interfere with the cellular composition of all worldly objects, including human bodies. I am seeking to find ways to express my own unexceptional humanness. I am an animal, an organic being with all its fragilities.

## Porosity

Frogs have a permeable skin, it makes them particularly vulnerable to chemical contamination, pesticides, herbicides, oil, heavy metals, and radioactive wastes, in the water, the air, and the soil. When the pH of creeks or ponds drops below 4.5, frogs disappear. "Frogs are an indicator species of toxic pollution—a kind of canary in the mines," proclaims my year 7 biology teacher. It is 1977; we are making our way through chapter one of the "Biological Science: The Web of Life." There is a picture of a pyramid, humans are at the apex—"humans are a complex intelligent social being," my teacher notes. All other living things are distributed below "they are 'simple' nature," he says. I was both fascinated and concerned; there was a storm water drain near my house where I would go to be with a host of frogs and others. A shimmer of oil sometimes glistened on the water surface in the late afternoon light, I worried toxins would be killing my storm water companions.

Scientists claimed humans were biological islands, (exceptional creatures) entirely capable of regulating their own internal workings. The specialized cells of our immune system taught themselves how to recognize and attack dangerous pathogens while at the same time mostly sparing our own tissues. Just as we have come to see we are not exempt from the Anthropocentric impacts we are having on the planetary systems, in recent times researchers have demonstrated that the human body is not such a neatly self-sufficient island after all. It is, like the planet, a complex ecosystem—an assemblage—containing trillions of bacteria and other microorganisms that inhabit our skin, mouth, and internal organs (Smith, 2015). In fact, most of the cells in the human body, my body, are not human at all. Bacterial cells in the human body for instance outnumber human cells ten to one (MacDougall, 2012). Haraway (2003) writes: "I love the fact that human genomes can be found on only about 10% of all the cells that occupy the mundane space I call my body … To be one is always to become with many" (pp. 3–4).

This mixed community of microbial cells and the genes they contain is collectively known as the microbiome (Ley, Peterson, & Gordon, 2006). All humans acquire this microbiome from very early in life, essentially during the birthing process and breastfeeding. Even though they do not start out with one, "Primate fetal development is thought to occur within an intrauterine microbiota-free environment, and yet within a short interval following birth the human microbiome is colonized" (Aagaard et al., 2012, p. 1). Each individual acquires their own community from the surrounding environment—our bodies therefore are an assemblage of nonhuman matter, the genealogy of our amorphous engagement with the ecosystem that makes up our being on and with the planet over our lifetime. I am a porous being.

## Lively matter

The wind that begins off the coast of Japan also typically travels eastward across the Pacific Ocean making its way to the west coast of North America, which is what happened immediately after the Fukushima Daiichi reactor explosions on March 11, 2011. "When those radioactive dust clouds turn into rain, the radionuclides become absorbed in the soil and, subsequently, the food chain" (Spector, 2012, p. 83). These invisible waves and particles of radiation flow like starlings in murmurations. I chose vital materiality as a theoretical space for exploring my data as it acknowledges the aliveness of matter—active, self-creative, productive, unpredictable. Matter as alive in entangled porous bodies is illustrated in this chapter theoretically through the story of radiation. I ascribe agency to this inorganic matter radiation to acknowledge it has a certain efficacy that defies human will "enchanted materiality" (Bennett, 2010). This enchanted materiality of radiation becomes entangled in the genealogy of my own being, and those bodies and things through which I share the planet. It is no longer a "human" body as an island but a posthuman assemblage of entangled things at a cellular level. The playgrounds in the cities near the disaster site still sit quiet and empty many years later (Kinoshita & Wolley, 2015). Children play inside. After the Fukushima meltdown, many parents were uncertain as to whether children should be forbidden from outdoor activities. Many parents restricted their children's outdoor play to reduce their exposure to radioactivity. Indoor play centers are booming in cities close to the nuclear site. In an old supermarket center in Koriyama city you can find a large public indoor activity space. The center supports 600 users and a variety of activities including a jogging track, a playhouse complete with large wet sand box, and a large variety of activity based equipment to support climbing, swinging, and jumping. Due to increasing demand, families are limited to 90 minutes per visit.

Children tell me they used to make daisy chains in the spring. After the Fukushima disaster the government clean-up teams stripped the outer bark off trees and removed the top several inches of soil. That kind of decontamination is specifically aimed at cesium, which falls out of the air like dust. It is transported by wind and clouds, then it washes out or it contacts and sticks to surfaces. If it falls on plants that animals eat, the animals get contaminated, too. On May 27, 2015 a Japanese citizen in Tochigi prefecture posted on Twitter the pictures of deformed plants in his neighborhood. Like the mutated insects after Chernobyl, these images conjure up monstrous matter. Swiss science illustrator Cornelia Hesse-Hoegger produced disturbingly beautiful watercolor paintings after collecting and documenting "morphologically disturbed" insects affected by the fallout blown from Chernobyl into Europe (Mok, 2011) (see www.smithsonianmag.com/arts-culture/cherno byls-bugs-art-and-science-life-after-nuclear-fallout-180951231/). After being criticized that the images were speculative fiction and were creating sensationalism, she documented insects from areas close to functioning nuclear power plants. Over 30% of these insects also had some deformity: misshapen wings, feelers, altered pigmentation, or tumors at about ten times the normal rates.

## Monstrosity

Jars of deformed dead bodies are contained in a local museum (*dark tourism*) in Semipalatinsk (see Figure 1.1). The children in the city I speak to tell me they have seen these "deformed bodies" penetrated by monstrous matter. Like the mutated insects and daisies they "provoke fear but also fascination as their ghostly presence, same but not quite threatens to reposition or dissolve the boundaries of normality" (Goodley, Runswick-Cole, & Liddiard, 2015, p. 3). It is in the invisible bodies of human and nonhuman beings that coexist in Semipalatinsk Nuclear Test site, in the steppes, or plateau region, in eastern Kazakhstan that my research studies on radiation and in particular a focus on childearthbodies, have been located.

**FIGURE 1.1** "Jars of deformed dead bodies." Dead foetus, congenital anomalies due to atom testing, Semey State Medical Academy
*Source*: Getty Images

During the cold war the Soviet Union chose eastern Kazakhstan as a nuclear testing site because it was one of its remotest, most desolate areas. From 1949 until 1989 Russia conducted 456 secret nuclear tests (116 above ground, the rest underground) at the site with a seemingly unfettered regard for the human and non-human bodies that coexisted there (the site is about 18,500 square km) (Keenan, 2013).

Over a 40-year period possibly as many as a million people, multitudes of birds, animals, fish, plants, water, the soil, the air were deliberately subjected to the impact of radiation exposure. No one was evacuated, nothing was excluded, with 1500 animals and thousands of local villagers placed strategically in the line of the fallout during each test. The ecological community, and in fact the planet itself, became a scientific experiment. Research by a Japanese team from 2002–2004 spoke to many villagers within 200km of the detonation sites. The outcome of their research was published in 2006:

> 90% of the respondents reported seeing the flash and 70% that they felt a bomb blast, 18% felt heat, 28% saw the mushroom cloud, 16% mentioned a "deafening roar", and 7% "referred to animals that had lost their hair", a typical sub-acute radiation injury.
>
> *(Kawano & Ohtaki, 2006).*

The villagers made it very clear to the researchers that they had not been informed about the nature of the tests, or of the dangers linked to them.

It was in 2014 that I travelled to Semipalatinsk nuclear test site (STS) in Kazakhstan for the first time. I flew 7000 kilometers by jet plane from Sydney to the capital Astana, and then onto a small Russian rotor plane to Uste-Kamenogorsk in the eastern region. Once

landed I then proceeded to travel the next 400 kilometers across the steppes to Semipalatinsk in an old Russian taxi (Figure 1.2). Around halfway along our journey the driver
pulled over to a dusty gas station to fill up with petrol. I decided to wander inside the
small shack on site to see if I could purchase cold water. This was to be my first encounter
with a dark ghostly stranger that traveled with me—invisible, pervading, deadly radiation.

Radiation is the emission of energy in the form of waves. The most penetrating form of
nuclear radiation are gamma rays. Radiation is invisible and not directly detectable by human
senses; as a result instruments such as Geiger counters are used to detect its presence. Inside the
small shack I encountered for the first time being tested for radiation with a Geiger counter. I
close my eyes—the sound reminds me of the photographs I have seen of children in Sendai,
Japan where I imagined these beeps as the apparatus ran across a small girl's body. The photograph
reveals the reading is low, her parents standing by are relieved. Radiation is monitored in the
streets, on the animals, on the plants, and in the air after the Fukushima meltdown. In the days
after the nuclear accident in Japan, the world's entire supply of Geiger counters was sold out.

I realize I am sweating as they wave the Geiger counter across my body. I conjure up
images: radioactive particles crashing through my body smashing into materials at such
speed they are colliding violently with atoms along the way—they are destroying delicately
balanced cells. They are penetrating deep inside my body causing fatal cancers to develop
or, if they infiltrate my reproductive cells, they are causing genetic defects. Once embedded
in the cell they may die, or they might just lie dormant, recovering only to support the
uncontrolled growth of cancer at some time in the future. The machine stops. I open my
eyes. The Kazakh man clad in his white body suit, Geiger counter in hand waves his hand
at me, to move on. I assume this means no radiation is detected. I don't buy any water; I
return to my taxi to continue my journey.

**FIGURE 1.2**  Taxi ride to Semipalatinsk, 2014
*Source*: Author

## Deadliness

Cancer rates were highest among residents of the villages and towns close to the test region (Figure 1.3 provides an image of the ferocity of the nuclear testing in the area). For many years during and after the nuclear tests, many of the local people complained of unexplained deaths and cancerous growths on them and their animal companions. Most of the birds, insects, and trees were lost. It became a dead, barren place. Life was tenuously hanging in the balance. Dirt and dust whipped around like great clouds of starlings, spread across the steppe, and settled across the landscape reaching the city of Semipalatinsk, which is a little less than 150 kilometers away.

> In an ecological study of childhood cancer incidence in four administrative divisions adjacent to the STS spanning from 1981 to 1990, an increase in relative risks for all cancers, including leukemia and brain tumors, were reported in children living less than 200 km from the test epicenter compared to children residing more than 400 km from the test site.
>
> *(Grosche, Zhunussova, Apsalikov, & Kesminiene, 2015, p.127)*

Plutonium, a heavy metal, emits alpha radiation; this material is most harmful when inhaled or ingested. On the testing site scientists find very high levels of plutonium in horse bones. Kazakh shepherds have returned to the site to herd their horses, leaving them to graze in the empty pastoral lands that were once the nuclear site. Kazakh shepherds use the bones of their horses to make soup. Horse soup and horse flesh are offered to me. I politely say no. I am a vegetarian.

We take a walking tour of the neighborhood (see Figure 1.4). Timur takes a photograph of a dead dog. As we pause, one child says to me, "I am afraid of the street dogs on the way home. Dead dogs stink." "Do you know about the nuclear tests?" they ask me as we walk. Yes, I say I did know. "It is inside us," one child remarks, "it is probably in you now" (Malone, 2018). My family and friends ask me why I would go to cities where there is evidence of radiation. I could answer with a complicated discussion of ethics, morality, children here having to live with it yet, whereas I am just passing through. But my answer

**FIGURE 1.3**   Archive image testing at the Semipalatinsk nuclear test site
*Source*: http://io9.com/5988266/the-tragic-story-of-the-semipalatinsk-nuclear-test-site

**FIGURE 1.4** Walking tour with children, Semipalatinsk, Kazakhstan.
*Source*: Author

**FIGURE1.5** Angelina's Dream dreaming
*Source*: Karen Malone

is simple: we are all entangled and implicated in an ecological posthuman community (Malone, 2018). There is no boundary for radiation, here or there, them or us—we are all fragile, porous beings, exposed to monstrous bodies.

Angelina shows me a photograph she has made during the winter, "I want a place where children can breathe fresh air. The factories make children sick in our city. Children get sent away." She draws me a picture of this dream place (see Figure 1.5):

> I love mountains because there is no mountains in our city. I love nature and animals but there is no nature or animals in our city. I would like to walk in the mountains. I want to take pictures of animals. I would want to explore the underwater world. And I would want to dance because of being happy to be breathing fresh air and away from the pollution.

## Assemblages

Monstrous bodies coexisting in Semipalatinsk are all in ecological "community." We are bound together in earth–body assemblages, a collection of genetically encoded messages and materials, passed on as radiation, reproduced in bodies, between bodies, and outside of bodies. The child/radiation/earth/bodies are not inert matter, but a community of entities exchanging through entangled relations cancerous potential. Through processes of territorialization, deterritoralization, and reterritorialization the assemblage can selectively be ordered, disordered, and reordered as newly reconstituted complex matters. Bodies are an assemblage of their traced histories as they act, react, and intra-act through time and space. In this posthuman entangled complex world, "[w]e do not leave our history behind but rather, like snails, carry it around with us in the segmented and enculturated installations of our pasts we call our bodies" (Hayles, 2003, p. 137). Children in the Anthropocene carry these material entanglements through their biome, in their stories, real and inscribed in the appropriation of what it means to cohabitate and reorder material "things" and "matter" in the damaged landscapes of their bodies and the earth.

## Anthropo-obscence

Within a short period after the accident in Fukushima the Ministry of the Environment reported high radioactive contamination in frogs species. In May 2012 within a 20-km radius of the Fukushima accident all species of frogs were extinct. Through posthumanist ecological narratives of child-contaminated-earth assemblages in Semey, through the concept of "porosity" I attend to the entanglement of all things, including the porousness of the unexceptional human species. The Anthropocene is not just a set of scientific facts, verifiable through stratigraphic or climatic analyses. Through a diffractive affective theorizing it comes to be a "discursive development," an unsettling ontology that problematizes a humanist narrative of progress that has essentially focused on the mastery of nature, domination of the biosphere, and the placing of a God-like faith in technocratic solutions. It is a heuristic device for gaining a deeper understanding of how we "humans" have come to locate ourselves as masters of a 4.5 billion-year-old planet when we have existed for the mere blink of an eyelid. I am encouraged through our storying of the Anthropocene to "track histories" to wander "through landscapes, where assemblages of the dead [and dying] gather together with the living" (Gan, Tsing, Swanson, & Bubandt, 2017, p. G6).

The term Anthropocene reminds us that traces of the past live on through those bodies who are amongst us—disasters and devastation formed our present—and that hope lies in considering these many pasts, as part of our future (Gan et al., 2017). The naming of the

Anthropocene produces opportunities to galvanize already emergent forms of thinking and acting in education. Changing the entrenched habits of modern western humanist thought, which is so adept at dividing humans off from nature, requires persistence, vigilance and a preparedness to take risks. It is hard work. It demands continually interrogating what it means to be human, to re-situate humans firmly within the environment, and to locate the environment within the ethical domains of a past and posthumanist landscape. Through, multi-matter storying of the Anthropocene, I am challenged to "track histories that make multispecies livability possible" by wandering "through landscapes, where assemblages of the dead gather together with the living"(Gan et al., 2017). . Traces of the past live on through those kin who are amongst us; disasters and devastation formed our present; and hope lies in considering these many pasts, as part of our future. Taking from Marisol de la Cadena (2015) who spoke in a conference I attended in Chile in October 2017, all bodies are "more than one—but less than many." The Anthropo-unseen/obscene, are the new assemblages of our catastrophic times. Sites of obscenity where we are "being worldly with" nonhuman kin in a landscape imbued with a fading past traced onto an uncertain present. These assemblages of fragile bodies, reaching out with threadlike tendrils encircling and clinging to an imagined shared future. "As humans reshape the landscape we forget what was there before … our newly shaped and ruined landscapes become the new reality. Admiring one landscape and its biological entanglements often entails forgetting many others" (Gan et al., 2017, p. G6).

By queering binaries and creating ontological openings, I am not offering up a flat ontology; rather earth/child/bodies is a movement, an encounter—not human, not radiation—child is human but not only, radiation is lively matter but not only. We are implicated in our existence on the planet with other matter despite the human predilection to reiterate human exceptionalism, including within many epic and heroic narrations of the Anthropocene. As humans, we are not exceptional or exempt, the Earth will continue on with or without us.

The children in Kazakhstan share their stories of childearthbodies penetrated by monstrous matter. Like the tumorous dogs, they "provoke fear but also fascination as their ghostly presence, same but not quite threatens to reposition or dissolve the boundaries of normality" (Goodley et al., 2015, p. 3). To attend, attune to, and be affected by the uneasy childearth encounters in the streets of Semipalatinsk is to recognize the porosity of bodies, radiation, lively matter, the catastrophic encounters that have been and continue to be the monsters of our contaminated toxic world.

## Openings

By altering climate, landscapes, and seascapes as well as flows of species, genes, energy, and materials, and a desire to continue on with a business-as-usual, we are sealing the fates of a myriad of species, in particular the future for our children. The proliferation of nuclear activities, a throwaway consumer society thanks in part to plastic, habitat destruction, accelerated species extinction, toxic pesticides, and air pollution have all heralded the beginning of a new age of the planet, the age of the human, the Anthropocene. Nowhere is this impact more vivid than the crowded cities of the majority of world nations. Cities that are microcosms of the planet fashioned for our [human] species and no other. Through my queering of human exceptionalism, I am creating ontological openings by reterritorializing assemblages of child/nonhumanbodies. Challenging the entrenched habits of modern western humanist thought, which are so adept at dividing humans off from nature, requires persistence, vigilance, and a preparedness to take risks. It is hard work. It demands continually interrogating what it means to be human, to re-situate humans firmly within the environment, and to locate the environment within the ethical domains of past, present, and future damaged landscapes. Posthumanism as a contemporary condition insists we think critically and creatively about who and what we are actually in the process of becoming—the new knowing subject—posthuman bodies are a complex assemblage that requires readjustments to our way of thinking. The Anthropocene no longer becomes a set of scientific facts, verifiable through stratigraphic or climatic analyses; radiation, air, water, earth become entangled with our humanness in unpredictable ways.

The urban environments of my research are central sites for troubling, reconfiguring, and reimagining human–nature encounters in times and spaces of the current planetary crises. Val Plumwood (2007, p. 1) wrote:

> The world is in ecological crisis and, if our species does not survive the crisis, it will probably be due to our failure to imagine and work out new ways to live with the earth, to rework ourselves and our high energy, high consumption, and hyper instrumental societies adaptively … . We will go onwards in a different mode of humanity, or not at all.

It is in the imaginary of a new way to be with the earth for our future children that my stories of the Anthropocene offer openings for speculating about a different future. And if our future has become entwined with that of the Earth's geological evolution, then contrary to the modernist faith, it can no longer be maintained that humans make their own history, for the stage on which we make it has now entered into the play as a dynamic and capricious force (Haraway, 2015, 2016).

Somehow we came to view ourselves as separate from nature. We have fallen into the trap that we are independent of the world—this is the human story we created. But the Earth is a living system; everything, including our children, is dependent on it for their survival. The naming of the Anthropocene can be utilized to do useful work, to galvanize already emergent forms of thinking and acting. This is the opening I offer to educators to embrace the opportunity that the Anthropocene presents as an ontological opening to provide a disruption of our everyday relations of being with the planet.

## References

Aagaard, K., Riehle, K., Ma, J., Segata, N., Mistretta, T.A., Coarfa, C., Raza, S., Rosenbaum, S., Van Den Veyver, I., Milosavljevic, A., Geyers, D., Huttenhower, C., Petrozeno, J., & Versalovic, J. (2012). A metagenomic approach to characterization of the vaginal microbiome signature in pregnancy. *PLoS One*, 7(6), e36466.

Barad, K. (2007). *Meeting the universe halfway: Quantum physics and the entanglement of matter and meaning.* Durham, NC/London: Duke University Press.

Barad, K. (2014). Diffracting diffraction: Cutting together-apart. *Parallax*, 20(3), 168–187.

Bennett, J. (2010). *Vibrant matter: A political ecology of things.* Durham, NC: Duke University Press.

Carson, R. (1962). *Silent spring.* London: Houghton Mifflin.

Crutzen, P.J., & Steffen, W. (2003). How long have we been in the Anthropocene Era? *Climatic Change*, 61(3), 251–257.

de la Cadena, M. (2015). *Earth beings: Ecologies of practice across Andean worlds.* Durham, NC: Duke Press.

Derrida, J. (2005). *On touching–Jean-Luc Nancy.* Stanford, CA: Stanford University Press.

Gan, E., Tsing, A., Swanson, H., & Bubandt, Nils. (2017). Introduction: Haunted landscapes of the Anthropocene. In A. Tsing, H. Swanson, E. Gan, & N. Bubandt (Eds.), *Arts of living on a damaged planet* (pp. G1–G14). Minneapolis, MN: University of Minnesota Press.

Goodley, D., Runswick-Cole, K., & Liddiard, K. (2015). The dishuman child. *Discourse: Studies in the Cultural Politics of Education, 37*(5), 770–784.

Grosche, B., Zhunussova, T., Apsalikov, K., & Kesminiene, A. (2015). Studies of health effects from nuclear testing near the Semipalatinsk Nuclear Test Site, Kazakhstan. *Central Asian Journal for Global Health, 4*(1), 127.

Haraway, D. (2003). *The companion species manifesto: Dogs, people and signicant otherness.* Chicago, IL: Prickly Paradigm Press.

Haraway, D. (2015). Anthropocene, capitalocene, plantationocene, chthulucene: Making kin. *Environmental Humanities, 6*(1), 159–165.

Haraway, D. (2016). *Staying with the trouble: Making kin in the Chthulucene.* Durham, NC: Duke University Press.

Hayles, N.K. (2003). Afterword: The human in the posthuman. *Cultural Critique, 53,* 134–137.

Kawano, N., & Ohtaki, M. (2006). Remarkable experiences of the nuclear tests in residents near the Semipalatinsk Nuclear Test Site: Analysis based on the questionnaire surveys. *Journal of Radiation Research, 47*(SupplementA), A199–A207.

Keenan, J. (2013, May 13). Kazakhstan's painful nuclear past looms large over its energy future. *The Atlantic.*

Kinoshita, I., & Wolley, H. (2015). Children's play environment after a disaster: The great East Japan earthquake. *Children, 2*(1), 39–62.

Ley, R.E., Peterson, D.A., & Gordon, J.I. (2006). Ecological and evolutionary forces shaping microbial diversity in the human intestine. *Cell, 124*(4), 837–848.

Lloro-Bidart, T. (2015). A political ecology of education in/for the Anthropocene. *Environment and Society: Advances in Research, 6*(1), 128–148.

MacDougall, R. (2012, June). *NIH human microbiome project defines normal bacterial makeup of the body.* Maryland, US: National Institutes of Health, US Department of Health and Human Services.

Malone, K. (2018). *Children in the Anthropocene: Rethinking sustainability and child friendliness in cities.* London: Palgrave Macmillan.

Massumi, B. (2015). *Politics of affect.* Cambridge, UK: Polity Press.

Mok, K. (2011). *Detailed illustrations of mutated insects challenge the science of nuclear power.* Retrieved from www.treehugger.com/energy-disasters/paintings-mutated-insects-nuclear-power-cornelia-hesse-honegger.html.

Nancy, J.L. (1991). *The inoperative community.* Minnesota, MN: University of Minnesota Press.

Plumwood, V. (2007). Journey to the heart of stone. In F. Becket & T. Gifford (Eds.), *Culture, Creativity and Environment: New Environmentalist* (pp. 17–35). Amsterdam: Rodopi.

Smith, A. (2015, September). *Small advances: Understanding the microbiome.* ABC Radio National. Retrieved from www.abc.net.au/radionational/programs/bodysphere/small-advances:-understanding-the-microbiome/6740394.

Smith, M. (2013). Ecological community, the sense of the world, and senseless extinction. *Environmental Humanities, 2,* 21–41.

Spector, H. (2012). 'Fukushima Daiichi: A never-ending story of pain or outrage?' *Transnational Curriculum Inquiry, 9*(1), pp 80–97.

# 2

# RENARRATIVIZING OUR EARTH-CENTEREDNESS

## A perspective from Aotearoa (New Zealand)

*Jenny Ritchie*

## Introduction

This chapter draws upon narrative research from Aotearoa that is informed by *te ao Māori*, a Māori worldview. These Māori onto-epistemologies inherently recognize and respect human relatedness and interdependence with(in) the physical and spiritual domains of *Papatūānuku*, the Earth Mother, and *Ranginui*, the Sky Father, along with their many children, the *Atua* (compartmental Gods, spiritual guardians) of domains including forests, people, oceans, winds, and cultivated foods. Because in this worldview people, along with other living creatures, are viewed as being descended from one of these Atua, Tāne Mahuta, we are closely related and spiritually connected to plants, insects, birds, and animals. Māori also include mountains, rivers, oceans, and trees as ancestors and spiritual guardians and thus care deeply and actively for these whanaunga (relatives).

Another aspect of te ao Māori is that elders are respected as the repositories of knowledge, the *kaitiaki* or guardians of the *tōhungatanga*, the collective wisdom. These knowledge guardians are also valued as role models, carers, and teachers of young children. This relationship is fundamental within the Kōhanga Reo movement that seeks to sustain *te reo Māori me ōna tikanga* (the Māori language and culture). A key Māori value to be sustained is that of *kaitiakitanga*, actively caring for the Papatūānuku, the Earth Mother, along with her mountains, rivers, and seas and all the creatures who rely on these for their wellbeing.

This chapter offers some examples of the dynamics of relationships between children, elders, and the earth in traditional Māori childrearing contexts as well as in more recent early childhood care and education paradigms. After drawing upon some narratives of traditional Māori childhoods and the significance of the bonds between *tūpuna* (elders) and *mokopuna* (grandchildren), the chapter offers a brief description of Te Kōhanga Reo, the Māori language revitalization and whanau development early childhood programs. Next, some examples are provided from the early childhood curriculum for Aotearoa (New Zealand), *Te Whāriki—He whāriki mātauranga mō ngā mokopuna o Aotearoa* (New Zealand Ministry of Education, 1996, 2017), which demonstrate the importance of relationships between children, ancestors, and our planet. The chapter will close with some narratives from research projects that illustrate the ways in which the knowledge of elders and respect for the Earth is valued and incorporated within early childhood care and education settings. We can treat this Planet Earth like an expendable machine, or we can recognize that our dignity is dependent upon the respect we pay to the origins of our earthly life and upon the responsibility we take for its preservation (Mead & Heyman, 1975, p. 197).

## Te ao Māori—the Māori world

Māori arrived in the islands of Aotearoa around the year 1200CE, and in the more than 500 years before the arrival of Europeans had developed their own unique onto-epistemology deeply grounded in their knowledge of the ecologies of their particular localities (Ritchie, 2013). The term "onto-epistemology" rejects the modernist division between knowledge and belief systems, recognizing the entanglement of ways of being, knowing, doing, and relating (Taylor, 2013). This knowledge had been adapted from their previous locales, which had been in much warmer, tropical climates. Many of the plants that they had brought with them to Aotearoa had failed to thrive in the more temperate conditions, necessitating a great deal of experimentation, intuition, and persistence in developing strategies for accessing resources for survival in the new land. These resources included food supplies, such as kaimoana (seafood), along with fruits, plants, birds from the *ngāhere* (bush, forest), and an extensive indigenous pharmacology (Tito et al., 2007; Waitangi Tribunal, 2011).

Māori onto-epistemology is deeply spiritual, with *wairuatanga*, spiritual interconnectedness, operating as an invisible web of inter-relationality between the tangible and intangible, the physical and the meta-physical. Furthermore, everything, every living or inanimate object, has its own spiritual force, or *mauri*. The concept of *mana wairua* requires that spiritual considerations are always present and need to be acknowledged (Ritchie, 1992). The Māori language demonstrates multiple, metaphorical levels of meaning, including spiritual inter-relationality of humans and the more-than-human realms. The word for land, *whenua*, is also the word for placenta. After the birth, the baby's whenua (placenta) and *pito* (umbilical cord) are returned to a special place, often a particular tree, on that child's ancestral whenua (land). These spiritual rituals acknowledge, celebrate, and solidify this special relationship of interconnectedness between people and their whenua. The Māori word that expresses their first nation's status is tangata whenua—people of the land. Thus, the extensive losses of Māori land as the direct result of settler actions has had a deeply wounding spiritual impact on Māori, as well as severe economic effects (Pihama et al., 2014). The loss of lands, language, and capacity to enact these spiritual rituals of wellbeing and healing have diminished Māori capacity to sustain their wellbeing. This has also been compounded by a racist education system that for many years denied Māori the right to an education in their own language or to the caliber of education that would equip them to access academic qualifications (Walker, 2004).

Māori philosophers and *tōhunga* (spiritual experts) such as Rangimārie Rose Pere (1982/1994, 1991), Hirini Moko Mead (2003), Māori Marsden (2003), and Hohepa Kereopa (Moon, 2003) have provided us with detailed explanations of Māori cosmologies with regard to inter-relatedness with earthly kin. These feature concepts such as *wairuatanga* (spiritual interconnectedness), *mauri* (life force present in both living as well as in seemingly inanimate things), and *kaitiakitanga* (active guardianship) (Marsden, 2003; Moon, 2003). Māori have consistently sought to have their traditional values and practices recognized by the wider society, and for restitution for the losses of land and language. A landmark report by a government tribunal recommended that *rongoā Māori* (traditional Māori healing using indigenous plant remedies) and its sources of vegetation should be supported by the Ministry of

Health and the Department of Conservation as one strategy for remedying ongoing poor Māori health outcomes (Waitangi Tribunal, 2011). Significantly, a recent landmark piece of legislation honored Māori worldviews by giving legal recognition to the Whanganui river as Te Awa Tūpua (The River Ancestor) (Office of Treaty Settlements, 2014; New Zealand Parliament, 2017).

## Role of elders in te ao Māori

Traditionally, elders had a key role in relation to child-rearing, which was an intergenerational, collective responsibility, shared by older siblings, both female and male, and elders of both genders. Renowned Māori anthropologist Te Rangi Hiroa/Sir Peter Buck explains that: "The etiquette of greeting and showing respect for one's elders was taught early" (Buck, 1950, p. 357). What Barbara Rogoff and colleagues (1995) have described as "guided participation" is the paradigm of early learning found in Indigenous and traditional societies, whereby children observe and participate in everyday culturally valued activities, in which the roles of elders feature prominently. For Māori, "[c]hildren sat on the outskirts of public receptions and social ceremonies and learned what was expected of them when they grew up" (Buck, 1950, p. 357). Te Rangi Hiroa explains further that:

> Much, if not most of the personal instruction in the early years, was received from grandparents ... The able-bodied parents were freed to devote full time and attention to the work which needed physical energy. The grandparents, who were too old for hard work, attended to the lighter tasks and the care of the grandchildren ... They told them stories and simple versions of various myths and legends ... The elements of a classical education in family and tribal history, mythology and folklore were thus imparted by male and female *tipuna [elders]* at an early age.
>
> *(Buck, 1950, p. 358)*

Mihipeka, who was born in 1918, describes how she received knowledge from her grandparents during her childhood, whilst they lived closely with the land and sea:

> There were times for stories, times for games, and that was usually in the early evening after everything was done. The koro [grandfathers] and the kui [grandmothers] used to sit down with the mokopuna [grandchildren]. That was their job, inserting knowledge into the children, and we were taught all sorts of things. That's how we learnt. That's how they passed the knowledge on to the children. They were the educators, the old people. The parents were busy getting food, they were strong and needed for the heavier work. Also they were absorbing knowledge themselves. Because you don't become professors in the Māori world until you are old.
>
> *(Edwards, 1990, p. 47)*

For Mihipeka, her elders were valued as guides and teachers who intentionally engaged children with learning to read the signs of nature:

> The old people were never put aside. They were the professors, te tohunga of Māori education in all fields. At a very tender age we would be taught about the sea and its many functions. We learnt to be in awe, to respect, to honour and to be very grateful. The sea yields good and healing for the body. We were taken to the sea to study

the many signs of nature applying to the very safety of humans going out to do deep-sea fishing, to watch the seabirds and to listen to their noises, to watch the clouds – if the sky is clear, you are quite safe; if the wind is more vigorous, make for the shore.

*(Edwards, 1990, p. 12)*

Also inculcated at an early age were the laws of *tapu* and *noa* (being sacred/spiritually protected or ordinary/free of restriction), which once understood and adhered to, provided spiritual safety and wellbeing. Mihipeka Edwards describes how during her childhood:

nothing was done without a karakia [spiritual incantation/prayer]. It was just a whole way of life in those days, to pray. My Kuia [grandmother] and all the other old people, they ... accepted their spirituality ... I would go with my Koro [Grandfather] into the bush and he would always karakia if he wanted to cut a tree down to use for making posts or things like that. Everything was precious to them, because those things came from the creation, the bush especially.

*(Edwards, 1990, p. 22)*

Children also became well versed in *rongoā*, the healing properties and uses of indigenous flora. Mihipeka Edwards explains that: "We lived by nature, and our medicines were all in the bush. There was a plant for everything – infections, infected sores, sterilising, stomach upsets, blood poisoning, poultices – everything" (Edwards, 1990, p. 17).

Venturing to the coast on an annual basis to gather seasonal foods, Mihipeka also learned from her kaumātua (elders) the spiritual protocols that demonstrated respect for the Atua, the spiritual guardians of the various domains, including bush and ocean:

On our first trip down to the sea, the first thing our old people would do was to karakia to Tangaroa (the Atua of the seas) to ask his permission to take food from the sea ... Kui (Grandmother) used to say we must respect all things in the sea because it is a privilege for us. And the sea water has many healing agents. If you have boils, open wounds, sores, muscular problems, rough skin the sea is the healer ... Down we went to the moana (ocean). The first sound I heard was my Kuia karanga [spiritual calling] to Tangaroa the atua of the sea.

*(Edwards, 1990, pp. 37–38)*

After the seafood was gathered, prepared and about to be served, a karakia of thanks was said before eating. Once the pots and dishes had been cleaned away by the older children, the *whānau* [extended family] would gather and again acknowledge Tangaroa, saying together:

| | |
|---|---|
| Ngā whakanui hoki | [Thanks and appreciation] |
| Kia koe te rangatira | [To you of high esteem] |
| Mō ēnei kai | [For this food] |
| Nau i hōmai | [You have given] |
| Hei oranga mō mātou | [To ensure our wellbeing] |
| Te atua o te moana | [The Atua of the sea] |

*(Edwards, 1990, pp. 39–40)*

Her grandparents also taught her about their ancestral taniwhā, Mukukai who resides in the ocean as a "wairua tiaki," or spiritual guardian.

With the coming of the British settlers, and despite promises to the contrary, the relentless process of colonization disregarded the spiritual guardians, desecrating sacred rivers, oceans, and lands in the service of profits and "progress." Mihipeka laments the desecration of the gifts of Ranginui, Papatūanuku, and the Atua:

> I realise now as I think back, that the seagulls, the many varieties, are the caretakers of the many shores of our beautiful land. Man has made so much work for them by letting body waste flow into the sea instead of turning it all into Papatūānuku to recycle. She, the mother earth, has her many creatures within her to do these particular kind of jobs. Instead, the Pākehā [people of British ancestry], with all his greed and arrogance and selfishness, thinks he knows everything. Turns all the filth like feces, urine, wastes from the factories and ships into the sea to kill the fish and shellfish. The energies from the salt water, I, the Māori, was taught, are a cure for many ailments. These are the gifts handed to us by the creator. Nā Ranginui anō i hōmai, mai noa atu ēnei taonga hei oranga mō tātou, aue te pouri e. [These are the gifts of Ranginui, the Sky Father, for our wellbeing, (a lamentation) of such sadness!— *translation by JR*]
>
> *(Edwards, 1990, p. 40)*

Mihipeka's grandparents' teachings enabled her to respect the birds of the bush and the wisdom that can be learnt from them. When Mihipeka asked her Koro [Grandfather]: "What is that little bird, Koro?" he replied:

> That is the grey warbler, the riroriro. She is a very aroha [loving] bird. This is the matua whangai [adoptive parent] for the pipiwharauroa [shining cuckoo] [who] pushes the riroriro eggs out [of the nest] and the poor riroriro ends up hatching the pipiwharauroa eggs. You see, Moko [grandchild], the birds teach us how to be kind, to share, to be just, to be aroha. They teach us to take good care of a manuhiri [guest], which are like the pipiwharauroa's eggs. This pipiwharauroa doesn't have time to hatch her eggs because she is too busy going from place to place to tell the many iwi [tribes] to hurry up and dig the soil, plant the kai [food] … But you see Nature takes care, because the riroriro have already hatched one lot of babies before pipiwharauroa comes home … from a faraway place for the springtime. It's not really a lazy bird. That bird goes round telling everybody, "Hurry up. Get up and dig the soil. Prepare. The spring has come."
>
> *(Edwards, 1990, pp. 58–59)*

Teachings such as the above demonstrate a view of interdependence, inter-relatedness, attunement, and respectfulness with/in the world of the bush and ocean.

## Te kōhanga reo

The Kōhanga Reo (literally, language nest) movement, is a national Māori immersion early childhood program focused on sustaining Māori language and traditions (Skerrett, 2007; Skerrett & Ritchie, 2016). Māori leaders initiated the movement in 1982 as a result of concerns discussed at a meeting of tribal elders in response to research that had demonstrated that very few members of the younger generations were speakers of the language, and that there were few domains remaining where the Māori language was the vehicle of communication (Benton, 1997). It was feared, therefore, that the language might die out with the passing of that generation of elders. The founding concept of the movement was to bring elders, fluent speakers of the Māori language, together with babies and young children, in an environment where only the Māori language was to be spoken. In this way it was intended to revitalize the language in a manner that honored the traditional role of elders/ grandparents in transmitting knowledge to their mokopuna (grandchildren). The emergence of the kōhanga reo movement precipitated a wider revolution of Māori reclaiming their rights to education that reflected their culture and identity (Smith, 2007), associated with ongoing Māori claims for settlement of historical grievances pertaining to the loss of lands, language, and traditional knowledges (Waitangi Tribunal, 2011). Te Kōhanga Reo is an expression of Māori tino rangatiratanga, the self-determination to live as Māori, via the medium of the language that articulates what it means to be Māori.

## Te whāriki 1996–2017

The first New Zealand early childhood curriculum (New Zealand Ministry of Education, 1996) was written during an era of increasing recognition of the protections that should have been accorded to Māori under the 1840 treaty, Te Tiriti o Waitangi, whereby Māori had legitimized British settlement. The four key writers of the document represented in-depth knowledge of the early childhood education field, on the part of Helen May and Margaret Carr, and deep knowledge of the Māori language and culture, held by Tilly and Tamati Reedy. An extensive range of different sector groups contributed to the document, as well as an extended consultation period. The document, both in its form, content, and development process represented a partnership that honored Māori as Indigenous peoples. Māori language and knowledges were expected to be highly visible in all early childhood services, although this posed an ongoing challenge to the early childhood workforce, which was largely monolingual in English.

Te Whāriki (NZ Ministry of Education, 1996) recognized the role of elders within Māori (and other) communities. For example, in explaining the principle of Families and Community | Whānau Tangata, it stated that: "Culturally appropriate ways of communicating should be fostered, and participation in the early childhood education programme by whānau [extended families], parents, extended family, and elders in the community should be encouraged" (p. 42). In the Strand of Contribution | Mana Tangata, the document elaborated on the importance of Māori and elders' contributions to the early childhood program:

> There should be a commitment to, and opportunities for, a Māori contribution to the programme. Adults working in the early childhood education setting should recognise the significance of whakapapa [genealogy], understand and respect the process of working as a whānau, and demonstrate respect for Māori elders.
>
> *(p. 64)*

In the Strand of Belonging | Mana Whenua, the link is made between Māori as tangata whenua [people of the land, Indigenous people] and Papatūānuku, the Earth Mother, pointing out the need for teachers to liaise with local tangata whenua and to promote respect for Papatūānuku (p. 54). Part B of the document, the Māori text, refers in depth to the importance of the child knowing their genealogy, and feeling their connection to Papa-tūānuku, the Earth Mother, Ranginui, the Sky Father, and to their children the Atua, the compartmental Gods, who are spiritual guardians of oceans, winds, forests, and cultivations. It reinforces the importance of raising mokopuna (grandchildren/children) to learn the stories of Papatūānuku and Ranginui, and their children, and to realize their role as kaitiaki (guardians) of the whenua (land) (New Zealand Ministry of Education, 1996).

The "refreshed" version of Te Whāriki (New Zealand Ministry of Education, 2017), contains a series of *whakataukī*, traditional Māori proverbs, threaded throughout the document to frame the introduction of each section. The whakataukī for the section on "Underpinning theories and approaches" is: *"Kia heke iho rā i ngā tūpuna, kātahi ka tika. If handed down by the ancestors, it would be correct"* (p. 60). The explanation provided is that "This whakataukī refers to intergenerational expertise and the respect Māori have for the wise counsel of the ancestors. It signals the importance of a credible, sound, theoretical foundation for teaching and learning" (p. 60).

## Elders integral to early childhood pedagogies: some examples from research in early childhood settings in Aotearoa

This section draws initially upon research previously published in a report to the funder, the New Zealand Teaching and Learning Research Initiative (TLRI) (Ritchie & Rau, 2006). In this project we found that for Māori parents and early childhood educators, a salient feature of their aspirations was the affirmation and valuing of traditional models of intergenerational transmission of language and culture and the valuing of the natural world. The role of kaumātua (elders) was significant, both in supporting the daily activities and as repositories of Māori language and the traditional knowledge that they were able to share with *tamariki* (children) and *whānau* (extended families). As one of the educators explained:

> I think our Nannies brought richness to our center. They just provided such awesome examples of tikanga [correct customs] and I knew that if I was unsure about something, I could just ask, and even at times they would predict and tell. And you need to be a humble person, but that was a really good experience for me to be able to step back and say: "Okay, no, I wasn't right, and this is a really good thing to be learning." [I really appreciated] their humour, even if the children weren't directly interacting with them, they could hear them nattering away [in the Māori language], laughing and that's just great because I can't provide that for my children and I really want that role model in my children's lives, having it within their early childhood [center] whenever we attended was just so precious.
>
> *(Sue)*

Another educator highly valued excursions that provided children and their families with experiences in the natural world, generating feelings of spiritual connectedness with the whenua (land):

> Oh, I think my ideal of a fully bicultural[1] Playcentre is that a lot of the time it wouldn't be at the center. We'd be out, we'd be out at the beach and sit in the rivers, doing the real stuff: eeling, cooking what you catch, looking after wherever you are. And I talk about as a child growing up and spending a lot of time at the beach and picking pipis and how we could ride our bikes around the streets. And, as

long as you turned up for your kai [food], life was sweet. So what do you want for your children? It's so much the same. I want my children to swim and dive and ride kayaks and ride their bikes and play on the farms and get out and about and learn all these things. So I think fully bicultural means there has to be a huge connection to this land. And looking after what we've got.

*(Miria)*

Similarly, for Ana, opportunities for working with natural materials, such as flax, provided a source of learning of traditional knowledge:

Harakeke [flax] became a vehicle to disseminate education about Māori values about our Atua Māori [Māori Gods], about a way to behave, tikanga [correct customs], ae [yes], everything. And our tamariki [children] learned alongside of us, we just provided opportunity for them too, they could do it just like us.

In a later project, also funded by the TLRI, which focused on "caring for ourselves, others and the environment" these themes of the involvement of elders in early childhood care and education settings along with the valuing of the natural world were also apparent (Ritchie et al., 2010). One kindergarten teacher discussed how they were seeking deeper understandings regarding Papatūānuku, in consultation with their local Māori community:

Papatūānuku is another real strength [in our] philosophy. That at the moment is in draft and we're discussing it because what does the wider concept of Papatūānuku [mean]; we could say Mother Earth but there's a wider concept to it and we need to work with all whānau and with our local iwi about what does that mean to them.

*(Head Teacher, Galbraith Kindergarten)*

Teachers from another kindergarten sought guidance from a local elder, Huata Holmes, along with their Senior Teacher, Lee Blackie, in utilizing knowledges particular to Huata's Southern Māori iwi (tribe), the local Indigenous people. The teachers made links to the early childhood curriculum, *Te Whāriki* (NZ Ministry of Education, 1996) in their report:

We consulted with Huata Holmes, our kaumātua, for guidance, expert knowledge and inspiration. The Southern Māori perspective or "flavour" is important. Lee Blackie, our Senior Teacher, accompanied Huata and gave us a practical aspect that could sit side by side with Huata's ideas. In order to add authenticity and depth, we arranged for Huata to come and narrate his southern mythology/stories/pūrākau to the children and whānau [families] (Communication/Mana Reo Goal 3: hear a wide range of stories, *Te Whāriki*, p. 59) as told to him as a child by his grandmothers and great grandmothers (Holistic Development/Kotahitanga: recognition of the significance and contribution of previous generations to the child's concept of self, *Te Whāriki*, p. 41). Huata's kōrero [narrative] was excellent and by working together we have achieved more of a shared understanding. He told of the great waka [canoe] of Aoraki coming through the sky down to the South Island. He also used the waiata [song] Hoea te Waka [row the canoe] to support his kōrero [talk]. This has become a real favorite. His kōrero has supported our teaching of the importance of Papatūānuku in our lives.

*(Richard, Hudson Kindergarten)*

In another kindergarten the teachers and children focused on recognizing the indigenous plants that were growing in the grounds, and had identified that some of these were used by Māori for therapeutic purposes, as rongoā, and that these rongoā were also linked to kai (food). This led them to locate a book that shared the stories of local Māori healers, *Matarakau. Ngā kōrero mō ngā rongoā o Taranaki. Healing stories of Taranaki* (Tito et al., 2007). The teachers talked with the children about the story of one local kuia (female elder), Kui Trish, who had grown up at Rangatapu, their local beach. They reported that:

> Kui Trish [relates how], she grew up at Rangatapu, Ohawe Beach. Life was simple and they were always well fed. The gathering and growing of kai [food] by her parents and whānau [extended family] was her rongoā [source of healing]. Their wellness as children was dependent on fresh air and food that had been either gathered from the sea or collected from the homegrown gardens. Kai was gathered and preserved. Everything was dried including dried whitebait and mussels. They made jams and pickles. She remembered her kuia walking down the hill with a kete kai moana [woven flax basket for seafood] to gather kai. They wove their kete with harakeke [flax] that grew nearby. Kai was gathered daily. Karakia [prayer] was said before gathering kai. They learnt the sounds of the sea, the signs of the incoming tide and the swiftness of the sea against the rocks. This was time to leave, even if their kete were not filled enough. They also gathered driftwood for the home fires. Whilst Kui Trish didn't have native bush around them, she acknowledged she has little knowledge about leaves used for rongoā. However, living at the beach provided all their rongoā.
>
> *(Hawera Kindergarten)*

According to this traditional wisdom, as passed down to children by the elders, from their elders in turn, living closely and respectfully with the land and sea was the source of wellbeing.

## Final thoughts

These brief excerpts from our previous research echo the stories told by Mihipeka of her childhood in the 1920s, of growing up under the watchful tutelage of her elders, participating alongside them in traditional practices of harvesting the fruits of forest and ocean, learning to accord due respect to the Atua (Gods) of these domains. Enabling children and their families to access these local histories and traditional knowledges via the wisdom of local Indigenous elders simultaneously provides authenticity as well as validating and affirming these knowledges (Ritchie, 2013, 2014).

Whilst *Te Whāriki* is an aspirational document, reviews of center practices indicate that considerable challenges exist for teachers who are not of Māori ancestry and experience in building the relationships that will offer them access to Māori onto-epistemologies (Education Review Office, 2012, 2013). The narrative assessment model promoted in early childhood care and education in Aotearoa requires that te ao Māori understandings be represented in pedagogical documentation, including a Māori view of the child, as portrayed in Te Whatu Pōkeka (New Zealand Ministry of Education, 2009). Previous research has demonstrated the willingness and commitment of some teachers within the early childhood sector to authentically incorporate *te reo me te ao Māori* (the Māori language and worldviews) within their programs (Ritchie & Rau, 2006, 2008, 2013; Ritchie et al., 2010). Yet, unconscious bias and the limitations of the current education system continue to work against this renarrativization (Blank, Houkamau, & Kingi, 2016).

The vast majority of New Zealand citizens do not speak Māori with any level of proficiency—the number of speakers in the previous census registered only 3.73% (Statistics New Zealand, 2013). *Te reo Māori*, the Māori language, encapsulates Māori ways of being, knowing, doing, and relating (Pere, 1991), and yet our national education system does not insist on te reo being systematically taught beyond the early childhood sector. Nor are Māori worldviews mandated as integral to the delivery of the primary and secondary school curriculum (New Zealand Ministry of Education, 2007). Teacher education is a core site for transforming the practice of teachers and ultimately the education sector (Ritchie, 2015). Providers of "mainstream/whitestream" teacher education qualifications are not required to set either entry or exit standards for their graduates with regard to competency in te reo Māori. Consequently, the teaching workforce is not proficient in te reo. The recent proliferation of one-year teaching qualifications indicates the need for urgent attention to these aspects, since it is difficult to acquire a sufficient level of proficiency in a second language in the space of one year whilst also covering all the other components of the teaching qualification. Currently, the New Zealand education system is undergoing a wide-ranging review, and it is to be hoped that these issues will be recognized and changes put in place to ensure that, in time, all citizens of Aotearoa New Zealand become proficient in the language and worldviews of Māori, the original people of this land.

## Note

1 'Bicultural' refers to the two main cultural groups of Aotearoa, Māori and Pākehā (of European ancestry).

## References

Benton, R. (1997). *The Maori language: Dying or eviving?* Retrieved from Wellington: NZCER. Retrieved from www.nzcer.org.nz/system/files/The_Maori_Language_dying_reviving.pdf.

Blank, A., Houkamau, C., & Kingi, H. (2016). *Unconscious bias and education. A comparative study of Māori and African American students.* Retrieved from www.oranui.co.nz/index.php/unconscious-bias.

Buck, T. R. H. P. (1950). *The coming of the Maori.* Wellington: Whitcomb & Tombs.

Education Review Office. (2012). *Partnership with whānau Māori in early childhood services.* Wellington: Education Review Office. Retrieved from www.ero.govt.nz/National-Reports/Partnership-with-Whanau-Maori-in-Early-Childhood-Services-Feb-2012.

Education Review Office. (2013). *Working with Te Whāriki.* Wellington: Education Review Office. Retrieved from www.ero.govt.nz/publications/working-with-te-whariki/.

Edwards, M. (1990). *Mihipeka: Early years.* Auckland: Penguin.

Marsden, M. (2003). *The woven universe: Selected writings of Rev. Māori Marsden.* T.A.C. Royal (Ed.). Wellington: The Estate of Māori Marsden.

Mead, H. M. (2003). *Tikanga Māori: Living by Māori values.* Wellington: Huia.

Mead, M., & Heyman, K. (1975). *World enough: Rethinking the future.* Boston, MA: Little, Brown & Company.

Moon, P. (2003). *Tohunga Hohepa Kereopa.* Auckland: David Ling Publishing.

New Zealand Ministry of Education. (1996). *Te Whāriki. He whāriki mātauranga mō ngā mokopuna o Aotearoa: Early childhood curriculum.* Wellington: Learning Media.

New Zealand Ministry of Education. (2007). *The New Zealand Curriculum for English-medium teaching and learning in years 1–13.* Wellington: Learning Media. Retrieved from http://nzcurriculum.tki.org.nz/.

New Zealand Ministry of Education. (2009). *Te Whatu Pōkeka. Kaupapa Māori assessment for learning. Early childhood exemplars.* Wellington Learning Media. Retrieved from www.education.govt.nz/assets/Documents/Early-Childhood/TeWhatuPokeka.pdf.

New Zealand Ministry of Education. (2017). *Te Whāriki. He whāriki mātauranga mō ngā mokopuna o Aotearoa: Early childhood curriculum.* Wellington: Author.

New Zealand Parliament. (2017). *Te Awa Tupua (Whanganui River Claims Settlement) Act 2017.* Retrieved from www.legislation.govt.nz/act/public/2017/0007/latest/DLM6831459.html.

Office of Treaty Settlements. (2014). *Ruruku Whakatupua. Te mana o te awa tupua.* Retrieved from www.wanganui.govt.nz/our-district/whanganui-river-settlement/Pages/default.aspx.

Pere, R. R. (1982/1994). *Ako. Concepts and learning in the Maori tradition.* Hamilton: Department of Sociology, University of Waikato. Reprinted by Te Kohanga Reo National Trust Board.

Pere, R. R. (1991). *Te Wheke.* Gisborne: Ao Ake.

Pihama, L., Reynolds, P., Smith, C., Reid, J., Smith, L.T., & Te Nana, R. (2014). Positioning historical trauma theory within Aotearoa New Zealand. *AlterNative: An International Journal of Indigenous Peoples, 10*(3), 248–262.

Ritchie, J. (2013). Indigenous onto-epistemologies and pedagogies of care and affect in Aotearoa. *Global Studies of Childhood, 3*(4), 395–406.

Ritchie, J. (2014). Learning from the wisdom of elders. In J. Davis & S. Elliot (Eds.), *Research in early childhood education for sustainability: International perspectives and provocations* (pp. 49–60). Abingdon, UK: Routledge.

Ritchie, J. (2015). Disentangling? Re-entanglement? Tackling the pervasiveness of colonialism in early childhood (teacher) education. In V. Pacini-Ketchabaw & A. Taylor (Eds.), *Unsettling the colonial places and spaces of early childhood education* (pp. 147–161). New York: Routledge.

Ritchie, J., Duhn, I., Rau, C., & Craw, J. (2010). *Titiro Whakamuri, Hoki Whakamua. We are the future, the present and the past: caring for self, others and the environment in early years' teaching and learning. Final Report for the Teaching and Learning Research Initiative.* Retrieved from Wellington: www.tlri.org.nz/tlri-research/research-completed/ece-sector/titiro-whakamuri-hoki-whakamua-we-are-future-present-and.

Ritchie, J., & Rau, C. (2006). *Whakawhanaungatanga. Partnerships in bicultural development in early childhood education. Final Report to the Teaching & Learning Research Initiative Project.* Retrieved from Wellington: www.tlri.org.nz/tlri-research/research-completed/ece-sector/whakawhanaungatanga%E2%80%94-partnerships-bicultural-development.

Ritchie, J., & Rau, C. (2008). *Te Puawaitanga – partnerships with tamariki and whānau in bicultural early childhood care and education. Final Report to the Teaching Learning Research Initiative.* Retrieved from Wellington: www.tlri.org.nz/tlri-research/research-completed/ece-sector/te-puawaitanga-partnerships-tamariki-and-wh%C4%81nau.

Ritchie, J., & Rau, C. (2013). Renarrativizing Indigenous rights-based provision within "mainstream" early childhood services. In B. B. Swadener, L. Lundy, J. Habashi, & N. Blanchet-Cohen (Eds.), *Children's rights and education: International perspectives* (pp. 133–149). New York: Peter Lang.

Ritchie, J. E. (1992). *Becoming Bicultural.* Wellington: Huia Publications.

Rogoff, B. (1995). Observing sociocultural activity on three planes: participatory appropriation, guided participation, and apprenticeship. In J. Wertsch, P. Del Rio, & A. Alvarez (Eds.), *Sociocultural studies of the mind* (pp. 139–164). New York: Cambridge University Press.

Skerrett, M. (2007). Kia Tū Heipū: Languages frame, focus and colour our worlds. *Childrenz Issues, 11* (1), 6–14.

Skerrett, M., & Ritchie, J. (2016). Kia tū taiea: honorer les liens. Confiance, éducation et autorité en Nouvelle-Zélande. Kia tū taiea: Honouring relationships. Trust, education and authority in New Zealand. *Revue internationale d'éducation de Sèvres, 72*, 103–113. Retrieved from http://ries.revues.org/5522.

Smith, L. T. (2007). The native and the neoliberal down under: Neoliberalism and "endangered authenticities." In M. de la Cadena & O. Starn (Eds.), *Indigenous experience today* (pp. 333–352). Oxford: Berg.

Statistics New Zealand. (2013). New Zealand Social Statistics. Māori Language Speakers. Retrieved from http://archive.stats.govt.nz/browse_for_stats/snapshots-of-nz/nz-social-indicators/Home/Culture%20and%20identity/maori-lang-speakers.aspx.

Taylor, A. (2013). Caterpillar childhoods: Engaging the otherwise worlds of Central Australian Aboriginal children. *Global Studies of Childhood*, *3*(4), 366–379.

Tito, J., Pihama, L., Reinfeld, M., & Singer, N. (Eds.) (2007). *Matarakau. Ngā kōrero mō ngā rongoā o Taranaki. Healing stories of Taranaki.* Taranaki: Karangaora.

Waitangi Tribunal. (2011). *Ko Aotearoa tēnei. A report into claims concerning New Zealand law and policy affecting Māori culture and identity. Wai 262. Te taumata tuarua. Volume 1.* Retrieved from www.justice.govt.nz/tribunals/waitangi-tribunal.

Walker, R. (2004). *Ka Whawhai Tonu Matou. Struggle without end* (revised ed.). Auckland: Penguin.

# 3

# ENVIRONMENTAL JUSTICE IN THE SHADOW OF THE HYPEROBJECT

## Reflections from (not) saving the community garden

*Casey Y. Myers*

"There is no 'pristine', no Nature, only history."
*Timothy Morton, Hyperobjects: Philosophy and Ecology After the End of the World.*
*Morton, 2013, p. 58.*

In this chapter, I retell two intersecting stories in which children's rights, environmental justice, and my own work as an activist play a role alongside town elders and contested city spaces. One story is about an urban garden lot and old house in a post-industrial city in the Midwestern United States and one is about the heavy metal element Lead. When these narratives collide, they make tangible the ways in which seemingly invisible forces operate, influencing each other over distances in time and space. In this time–space of collision, a *hyperobject* (Morton, 2013) emerges and so many of the deeply held notions of the world and our place in it are displaced. This reordering of time and space causes me to reexamine my notions of the ways in which environmental justice materializes for/with children; it leaves me in a space of questioning where and how (and even when!) to move next.

## What is the hyperobject?

Before I begin to tell these stories, it will be helpful to the reader to gain an understanding of hyperobjects. Termed by philosopher Timothy Morton, hyperobjects describe phenomena in the *Anthropocene*—the current epoch that began when human activity emerged as a geological force. Examples of hyperobjects are all around us: global warming, the biosphere, capitalism, plastics, fossil fuels. Within our biosphere Earth, hyperobjects are inescapable in that there's no way to get out of them or get them off of you (e.g., they are *viscous*). Hyperobjects are everywhere and nowhere in particular. We feel their effects, but are incapable of determining their boundaries (i.e., they are *nonlocal*). Hyperobjects are so massively distributed across time and space that we can only ever experience a small fraction of the phenomena during our lifetimes (i.e., they are *phased*). Hyperobjects are made up of many parts but can't be reduced to one "thing" or another (i.e., they are *interobjective*) (Morton, 2013).

Because of these characteristics, our engagements with hyperobjects are "fascinating, disturbing, problematic, and wondrous" (Morton, 2013, p. 62). They bring about what Morton calls *the end of the world*. The Earth is still here—itself a hyperobject —but what Morton means by the "world ending" is that when hyperobjects make themselves known, our inherited ways of conceptualizing the world are no longer sufficient. Hyperobjects

> cause us to reflect on our very place on Earth and in the cosmos. Perhaps this is the most fundamental issue – hyperobjects seem to force something on us, something that affects some core ideas of what it means to exist, what Earth is, what society is.
>
> *(p. 24)*

Hyperobjects humiliate us, they reveal that "all entities are fragile ... and hyperobjects make this fragility conspicuous" (p. 13). They rearrange our world/view.

## The Garden and the very old house

In the North end of the downtown of our city, there was a small, sloping lot that existed in a quasi-feral state for over two decades (Figure 3.1). An Arts Organization that was housed in the building next door to the lot tended to it. They occasionally mowed the grass and began planting native species of plants and trees. As the Arts Organization grew and the community of artists formed around it, the coordinators of the organization extended and deepened their involvement with/in the plot of land. They began to plant large beds of garlic and kale that were free to anyone who may have been passing by. As the Arts Organization expanded to house a children's theater program, a rope swing and wooden platform appeared on the yard, along with painted rocks and plastic hula hoops, and an occasional deflated soccer ball. A small fire pit was made at the top of the lot near

**FIGURE 3.1**   The Garden lot, looking from the top of the hill onto the street below
*Source*: Casey Y. Myers

the kale beds and was sometimes surrounded by a few folding metal chairs or cinder blocks or mismatched pieces of plastic lawn furniture. Children painted sets for their plays along the grassy hill and ran into the yard to pick apples while their parents ran errands or had meetings downtown. Children were free to come and go as they pleased and were often unsupervised in this space. Although the people associated with the Arts Organization never paid any rent on this lot or asked permission to use it, their involvement continued in this way for many years. Among those familiar with the Arts Organization and/or the children's theater program, this space was casually referred to as "the Garden."

Several blocks in the other direction, there was a Very Old House (Figure 3.2)—it was built in the mid-19th century, prior to the American Civil War. This house once belonged to the daughter of the founder of our town. For the past 50 years it had been used as a boarding house for college students and though it still retained some of the original architecture, the university did not consider it worth saving when they drafted plans for the new esplanade that would connect campus to the downtown area. The Very Old House was on a parcel of land that the university wanted and so they were going to tear it down. A local group of preservationists formed a non-profit and rallied to save the Very Old House. This organization was permitted to buy this house for one dollar on the condition that it would be moved in time for the University to continue its construction on the esplanade. Despite the low sale price, the organization didn't have enough money to move and restore the house, so the City Council decided to loan them 15,000 dollars. Its proposed location was the only undeveloped parcel of land in our newly revitalized downtown—the aforementioned "garden."

During this time, I joined the core group of organizers working to "Save the Garden." My role was to promote it as a children's space. I spoke at several city council meetings. I interviewed the children who frequented the yard and worked as a documentarian. I took their words to the council and urged them to listen to their youngest citizens.

> "Why doesn't anything belong to kids?
> We have ideas.
> We have good ideas about this world!"
> Ilana, age 12

> "I know all about this garden and how to find the wild strawberries to eat.
> Grown-ups won't even try them.
> But they want to tear down the garden."
> Lester, age 5

I strategically tried to appeal to neoliberal ideals of growth and development; in speeches to city council, I cited research that said that these kinds of greenspaces improved children's cognitive skills (e.g., Burdette & Whitaker, 2005; Taylor & Kuo, 2011), and cited economic statistics that spoke to the ways in which this non-profit Arts Organization contributed to the city's wellbeing and argued that this lot was essential to the organization's functioning. I also made appeals to their values, urging them to consider the fact that building community does not require *buildings*. (I thought this was a particularly clever turn of phrase!) I pressed city council members to answer publicly whose history matters more: a long dead daughter of our city's founder or the children right in front of us? As one might expect, I was never answered directly.

**FIGURE 3.2**  The Very Old House
*Source*: Casey Y. Myers

Community tensions were high at this time. Many people who had formerly been allies—those concerned with historical preservation and restoration and those who worked in the arts community—were choosing sides. Hand-drawn "Save the Garden" signs began to appear in the windows or on the noticeboards of local shops (see Figure 3.3). "Our side" out-numbered "their side" 10-to-1 during the public hearings. When it came to light that the city council had violated the law when they pushed through site plans for placing the house on the lot, a local lawyer took on our case pro-bono and we filed a lawsuit against the city.

Much of this "good" activist work involved strategic moves to shape the discourse surrounding this plot of land. Others called the space an abandoned lot, we only referred to it as "the Garden." The lot was elevated to a symbol, the meaning of which depended on who was doing the interpretation. For supporters of the Garden, it became a symbol of the bottom-up community building that couldn't simply be replicated somewhere else. It was also seen as symptomatic of the ways in which the so-called revitalization of our city was really a top-down whitewashing, a gentrification process that was squeezing out the last spaces that long-time residents and their children enjoyed. For supporters of the Very Old House, the Garden was a symbol of the city's need for revitalization and restoration. It was a symbol of modern urban decay. The swing and campfire ring were simply zoning violations (*they were*). They claimed the sloping lot was "too steep" for children to play safely (*it wasn't*) and this posed a danger, considering small children might somehow roll right into the traffic below. They claimed that our side was reckless—children had no business playing in an abandoned lot alongside not only raspberries and apples, but also scrap metal and empty beer bottles. The lot was framed as an eyesore that attracted unsavory people that loitered, keeping the "culture" of the neighborhood from being "restored."

These arguments that framed the Garden as a danger or a blight were steeped in race, class, and age prejudices. Many of the folks on "our side" were working-class single parents, more racially diverse as a group than our city at-large, underemployed artists, and people with tattoos and visible disabilities. College students. Children. The folks on "their

**FIGURE 3.3** Hand-drawn sign of support for the Garden in a local restaurant
*Source*: Casey Y. Myers

side" were White, upper-middle class, middle-aged. One of the children in our organizing group pointed out these disparities to me as we stood in the parking lot of city hall after a contentious July council meeting.

> "When people look like them, they always win.
> They look like they never sweat."
> Larissa, age 11

And, she was right. After over a year of organizing and protesting and legal battles on both sides, the "other side" did win. The city approved their loan, the Garden was bull-dozed and excavated. The Very Old House was moved to the site (see Figure 3.4).

## The lead paint

The second story is shorter, at least the way I'm going to retell it. But it's actually longer than what I'm capable of thinking. And as it accelerates, it seems to swallow the previous

**FIGURE 3.4**   The Very Old House at its new location (the former Garden)
*Source*: Casey Y. Myers

story. It's a story about lead. Lead is easily mined from the Earth's crust and refined. Lead is soft, malleable, dense, and resistant to corrosion. Lead pipes carried our water, fortified our gasoline, and gave our paints lasting power. In the United States, lead was phased out of commercial and household products in the 1970s and '80s because it's also a neurotoxin that can be absorbed through the skin, ingested, or inhaled.

Unlike some other potentially toxic substances, there is no known safe level of lead—there is no lower threshold that is actually "okay" for us. Leaded gasoline and most leaded paints have been phased out of commerce, but they never truly go away. Lead is coursing in our blood and deposited in our bones, infiltrating our body systems via our organs. Our city's place in the *rustbelt*—a swath of postindustrial Midwestern territory that bustled from the 1940s to the early 1980s—seats us firmly within leaded land. We literally live in lead. Lead is us. Some of us are more lead than others. Children are more likely to be lead than adults, simply because they are close to the ground where the heavy lead falls and stays. The number of children qualifying as "poisoned" in the United States exceeds 1.2 million (Roberts et al., 2017). Symptoms of high lead levels can include lethargy, organ failure, impaired executive functioning and cognition, undifferentiated global delays in development, and death (Centers for Disease Control, 2018).

## Hyperobject emerges

The intersection of lead and this garden lot and this house and children's rights to a physical place is where/when the hyperobject emerges. Because of all of the characteristics listed previously—viscosity, nonlocality, phasing, and interobjectivity—hyperobjects often only come into view during a trauma that reaches across the false binary between humans and nature. When they do, "a deep shuddering of temporality occurs" (Morton, 2013, p. 25). In this particular case, when the dirt was exposed to excavate the lot and the house's paint was scraped during renovation, the hyperobject lead was there. It had been looming the whole time, working across time–space, unfazed by our activism or our public discourses.

A year after we had lost our battle to save the Garden, I was working as an art director for the children's theater program, which was still operating in the building next door to the newly placed house. I was walking some of the children to the coffee shop next door to meet their families after play practice. Although their city permits stated that proper precautions were being taken, I watched as over a century of paint was sanded from the siding of the house, throwing clouds of dust that undoubtedly contained lead into the air. My reaction was to grab the children and run away, but as lead is a heavy metal, we didn't have the time. It would have immediately fallen on top of us and back into the freshly excavated soil.

In our plans to work with the children to process what had happened to their Garden, one of the ideas put forth by the children was for us to make amends and work together with this Very Old House preservation organization so that they could potentially develop a set of joint bylaws by which they could still have rights to use the land that remained *around* the house for gardening and playing. But we could no longer physically do our justice work on-site. Any ideas or intentions for what might be possible in that place for children in the future were dramatically rearranged once that paint was exposed. And currently I'm still in this space of wondering how we can maintain community connections in light of all of these literally and figuratively toxic relationships (Figure 3.5).

## What now?

The fact that lead never entered into our debate about children's rights to this land is emblematic of the ways in which "we humans are playing catch-up with reality" (p. 21). The hyperobject lead was *always* already here (there, everywhere). Lead exerts its force and will continue to do so, perhaps modified by our human interventions but never going away. As Morton (2013) states, "knowledge of the hyperobject Earth, and of the hyperobject biosphere, presents us with viscous surfaces from which nothing can be forcibly peeled. There is no Away on this surface, no here and no there" (p. 38).

In my role as an activist on behalf of children, I was fixated on changing the *discourse* and I believed that justice on behalf of children depended on one discourse prevailing—either the garden would be valued as a children's space and a vital community epicenter or devalued as an abandoned blight or nuisance property that needed revitalization. And as a means to this end, I was willing to take up neoliberal discourses of persuasion that I didn't even agree with. I tried to move *others* to change sides by speaking their language of "return on investment." I made pleas to the benefits of a pure and restorative "Nature" that I don't even believe in. But in the face of the hyperobject lead, I have come to realize that "this taking of 'sides' correlates all meaning and agency to the human realm, while in reality it isn't a question of sides, but of real entities and human reactions to them" (Morton, 2013, p. 100).

The workings of environmental (in)justice are complex. The emergence of the hyperobject tells me that they may be more complex than I am capable of thinking. Many of us who have been engaged in activist work on children's rights already understand that systems are connected—justice for young children is connected to race and class and gender and place. But the boundaries of these human systems need to be renegotiated so that we might be able to grapple with the end of our world as we know it. It has been argued that in this posthuman moment "we need new relationships, and urgently" (Dahlberg & Moss, 2013, p. xii), that the larger web of relations in which childhood, and indeed all of us, is

**FIGURE 3.5**   The Hyperobject emerges at the intersection of these two landscapes

*Source*: Casey Y. Myers

entangled must be remade. But, in the shadow of these hyperobjects, the more urgent need seems to be comprehending what relationships both preexist *and* outlast us. How might we be humbled into recognizing that "non-human beings are responsible for the next moment of human history and thinking" (Morton, 2013, p.188)? How might justice with/for children come to encompass the multi-temporal, the multi-scalar, the super-human, the infinite?

## References

Burdette, H. L., & Whitaker, R. C. (2005). Resurrecting free play in young children: Looking beyond fitness and fatness to attention, affiliation, and affect. *Archives of Pediatric and Adolescent Medicine*, *159*(1), 46–50.

Centers for Disease Control and Prevention. (2018, October, 18). *Lead*. Retrieved from www.cdc.gov/nceh/lead/default.htm.

Dahlberg, G., & Moss, P. (2013). Introduction by the series editors. In A. Taylor (ed.), *Reconfiguring the Natures of Childhood* (pp. ix–xiii). New York: Routledge.

Morton, T. (2013). *Hyperobjects: Philosophy and ecology after the end of the world*. Minneapolis, MN: University of Minnesota Press.

Roberts, E. M., Madrigal, D., Valle, J., King, G., & Kite L. (2017). Accessing child lead poisoning case ascertainment in the United States, 1999–2010. *Pediatrics, 139*(5), 1–8.

Taylor, A. F., & Kuo, F. E. (2011). Could exposure to everyday green spaces help treat ADHD? Evidence from children's play settings. *Applied Psychology: Health and Well-Being, 3*(3), 281–303.

# PART II

# Earth-indigeneity

Place and pedagogies

# 4

# PLACE SENSITIVE PEDAGOGY AND THE IMPORTANCE OF TRADITIONAL KNOWLEDGES IN SÁMI EARLY CHILDHOOD INSTITUTIONS

*Aslaug Andreassen Becher, Laila Aleksandersen Nutti and Bushra Fatima Syed*

## Introduction

This chapter is focusing on Sámi early childhood institutions in the city of Oslo, Norway, and in rural settings in the northern part of Sápmi. Sápmi is the geographic area stretching across the countries of Norway, Sweden, Finland, and Russia—the historical settlement of Sámi people who are indigenous to this area. The Sámi people have been living in this area since prehistoric time, and languages, traditions, crafts, and ways of life can be traced back 2000 years. Here we draw on elders' interviews in order to discuss how *place*, the individual, and pedagogy are positioned in relationship to one another, in order to search for valuable practices in a worldly/global early childhood perspective. Through conversational interviews with elders, educators and resource persons, the concept of pedagogic and natural settings for institutions as arenas for preserving and revitalizing cultural knowledge has contributed to our reflections on both indigenous and common early-childhood education. The term "indigenous" refers to an internationally recognized identity for native peoples that emerged in the mid-1970s (Kuokkanen, 2007). Indigenous peoples, characterized as first peoples, share certain experiences of colonialism, as well as values and worldviews based on a holistic and close relationship to the natural environment (Kuokkanen, 2000; Smith, 2012).

The rest of the chapter is structured in four parts: First, we offer a theoretical summary of *place* referring to different understandings in the late 20th century, supported by the ideas of Agnew (1987) and Massey (2005), among others. Additionally, we consider how Massey's concept of place can inspire or offer pedagogy embedded in indigenous perspectives and backgrounds. Massey (2005) draws on the philosophy of Deleuze and Guattari, among others, when she discusses spatiality and argues for the multiplicities and openness of concepts. "Deleuze and Guattari, for instance, argue that a concept should express an event, a happening, rather than a de-temporalised essence," (Massey, 2005, p. 28). Similarly, we find this inherent in Massey's concept of "place." In our study, we detail the methodological landscape of our research, drawing upon Law (2004) who, like Massey, sees methods as performative, as part of the construction of "place," as well as the experience and interpretation of place. In the third part of the chapter, we focus on the Sámi kindergarten practices—specifically the place-related thinking foregrounded through the conversations with our informants. Taking our theoretical perspectives into account, we finally discuss the relevance of our research within Sámi institutional settings for practices and pedagogies in early childhood curricula.

## Approaches to place as a concept

The word "place" hides many differences, says Tim Cresswell (2004). In describing different approaches that have been articulated as "place," we follow Agnew (1987), who identified three fundamental aspects of place: *location, sense of place*, and *locale*. Even if Agnew's (1987) major theme is to argue that political behavior is intrinsically geographical, his thorough work and categorization are followed by geographers around the world (e.g. Cresswell, 2004; Førde, Kramvig, Berg & Dale, 2013). Theorists argue there are not strict borders between the approaches to "place" in the literature, but focus and attention has been and still is upon different understandings to support theorizing human relationships within place. In the following description of our understandings of place-based research, we also lean on human geographers and their interpretation of Agnew (1987). Førde et al. (2013), for example, hold that three main understandings within place-based theories are: complementary, intrinsic, and mobile boundaries. "Place" also encompasses the dimensions of materiality, everyday life, and experience.

*Place as location* would stress material and physical dimensions of place. People and groups of people are given characteristics/properties in a static way. However, the static perception of place as location would still have a significant focus in present time (Førde et al., 2013). People and places are described in research as entities with clear borders, origins, and identities. We are able to "locate" the place on the map, and the material physical dimensions such as buildings, roads, parks, etc. are described. Agnew (1987) says about *place as location*: "This involves not only everyday social practices but the long-run siting of locales through the distribution of resources and the physical construction of settings" (p. 26).

*Sense of place* as *genius loci*, the second approach, was developed by classical geographers. In addition to being located, *places* have a relation to people and/or people have a relation to places. This relation is subjective and emotional (Cresswell, 2004). Places produce meaning, and researchers claim it possible to find an inherent, authentic, and true meaning of place. Place-based approaches, when used in conservative ways, have been considered excluding, reactionary, and even racist, with some people or groups of people "belonging" and others not (Holloway & Hubbard, 2001). A political issue in present European politics, for example, is related to refugees and immigrants and whether they have the right to belong and have a proper place in Europe. However, theoretical contributions from Castree (2009) and Massey (1994) underline that *sense of place* is not about a unique relation between a person and location, but is a general sense that there are multiple variations on feelings of belonging and the ways people experience place. Within place as genius loci, a subjective sense of place is not necessarily static or permanent, pleasant or unpleasant, but always in motion and dynamic. Even if identities are based in place, they are not necessarily bound to place (Castree, 2009). Both sense of place and identity of place (genius loci) are ambiguous and influenced by political or global issues. A "progressive sense of place" implies that we have tolerance toward cultural differences and move away from mental and social barriers around "our place" as a possession only we hold, and toward one permeable to the needs of others as well as ourselves (Massey in Førde et al., 2013).

*Place as locale* is the last of the three conceptualizations of place. Cresswell (2004) draws from geographers who put forward new understandings focusing more on humans' "everyday lives," the body in movement and place as embodied and *becoming*. Relationships between overarching structures, such as worldwide economic capitalism, also locate social structures and our ability to exercise agency in our everyday lives. Such structural patterns have also informed some of these "new" approaches to place (Cresswell, 2004, p. 35).

The work of Doreen Massey has been most influential in reconceptualizing *place as locale* by featuring the socio-material, gendered *and* spatial contexts for everyday life, as they occur in face-to-face-interactions (Førde et al., 2013). This understanding of place is *relational* and has included global and international structures in combination with everyday political events.

## Place as locale—related to barnehage (kindergarten)

A *barnehage* in Norway would not be a kindergarten without children playing and learning, toys, children's books, and the materiality related to small persons. Children and professionals are conforming to expectations of what one does in an early childhood institution. Individuals are producing and are being produced by daily events in the *place* of the school. The "production" can change, however, by moving the person's body and the expectations beyond the fences of schools and into nature, where playing and learning, with children and professionals, continues.

Place as locale can be applied to early childhood, but is now mainly embraced in the world of geographers, anthropologists, and sociologists. Massey (2005) has argued that places are like networks and meeting places of the material, where social and cultural relations stretch beyond the locale. She talks about the fluidity and "thrown-togetherness" of places and argues that places are also events. *Place* is not stable or closed; it is open and changeable, even as nature is continually changing. Massey's (1994) notion of place is "an endless unfolding of difference," and it encourages a consideration of places as progress. Cresswell (2004) draws on the work of Lefebre and de Certeau, among others, when he also summarizes this third approach, *place as locale*, as "an event rather than a secure ontological thing rooted in notions of the authentic. Place as an event is marked by openness and change rather than boundedness and permanence," (Cresswell, 2004, p. 39).

An understanding of place(s) as relational can point to place(s) as assemblages not temporally fixed, and where the vitalism of the assemblage also refers to the relationships between and beyond materialities. Arguments for focusing on the more-than-human aspects of social relations on earth are of political importance to the environment. Deleuze and Guattari argue this when they suggest that "earth" should be introduced as a category (May, 2005). Their philosophy focuses on life as an infinite plane, without considerable divisions between various elements on earth.

> Climate, wind, season, hour are not of another nature than the things, animals, or people that populate them, follow them, sleep and awaken within them ... This animal is this place!
>
> *(Deleuze & Guattari, 1987/2004, p. 290)*

Duhn (2012), who seems to appreciate these perspectives, holds that early childhood still suffers from a general lack of awareness of how place as force and form continuously influences the self in its relationship to the world and the "earth." If we follow assemblage as a concept of relevance in place-based pedagogy (whether location, relation, or locale), we can spot the vitality of materials and elements that together *with humans* constitutes, or makes up, place, Duhn argues.

## Indigenous knowledges in pedagogies

Both Sámi and other indigenous researchers emphasize the importance of incorporating indigenous peoples' worldviews, values, traditions, and ways of learning into indigenous education through decolonizing processes (Greenwood & de Leeuw, 2007; Hirvonen & Balto, 2008; Y.J. Nutti, 2011b; Rameka, 2012, 2016; Smith, 2004, 2012). Decolonizing and transforming processes are used to let traditional knowledge be part of the living curriculum. Balto (2008) and Bergstrøm (2001) write about how traditional knowledge forms a common understanding and insight, and creates tools for self-articulation. The holistic aspect of traditional knowledge, connection to nature, and its spiritual dimensions is important within the framework of education (Aikio, 2010; Guttorm, 2011; Hirvonen, 2009).

Furthermore, Kincheloe and Steinberg (2008) write about the transformational power of indigenous knowledge, the way such knowledge can be used, the benefits, and how to ensure that the interests of indigenous people are served. Several works (Greenwood & de Leeuw, 2007; Rowan, 2017) argue for place-connected pedagogies and holistic education, where land, health, and well-being are related to the education of young indigenous children. Additionally, L.A. Nutti (2011a) and Balto (2008) write about how to transform and to decolonize processes in education, and how such techniques can change the way of thinking and acting in one's professional role as educator.

Over time, the contexts where Sámi children acquire knowledge have changed sharply, and both families and educational institutions have new roles and responsibility for using and developing traditional Sámi knowledge and practice (Balto, 1997; Becher, 2006; Nergård, 2005). Sámi early childhood institutions across Sápmi are an important arena for transforming and developing newer ways of using traditional knowledge. These institutions are also found in urban settings outside the main Sámi areas, and throughout Sápmi they have various national frameworks and curricula (Nutti, 2011, 2014), because Sápmi consists of several nations (Norway, Sweden, Finland, and Russia). The indigenous content is local-, place-, and culture-based, and it is the desire of educational leaders to transmit Sámi and indigenous knowledge and traditions more fluidly into the Sámi early childhood education curriculum (Balto, 2008; Nutti, 2011, 2014; Storjord, 2008). Teachers require an intense awareness of the need to transmit traditional Sámi knowledge to children, because Sámi perspectives and contents also mean a strong affiliation/connection to place. Nutti and Joks (2018) write about the significance of *places* for the early childhood institutions through traditional Sámi practices, discussing, for example, the tradition of marking reindeer calves as one instance of place-based practice.

### *Learning from indigenous elders and others*

On discussing the concept of "place," we asked our informants whether belonging to place is about belonging to *places*. The informants (Nos. 1 and 2) answered yes to this question, reminding us that moving with the reindeer is a normal part of nomadic life. Additionally, the Sápmi is an area that comprises four countries: Norway, Finland, Sweden, and Russia. So while our national identity might be considered Norwegian, a more productive place-based understanding, which considers the relationships between place, people, and things, is a more reasonable assemblage for children and teachers in terms of learning.

Place and belonging is fluid, right? But at the same time, very concrete because of the relation between human and nature. To talk about several places [for belonging] is very central.

*(Inf. 1)*

## Connecting "place" and "place pedagogies"

Are there possible connections between conceptualizations of place and how educators should/can act in professional relationships with children? Commentators who advocate place as relevant in educational settings include Somerville (2007, 2010), Ritchie (2012), Duhn (2012), and Melhus (2012). Furthermore, Somerville (2010) explains how she wanted to explore these ideas and enable an articulation of place-based pedagogy through the lenses of poststructural feminist theories. She explored what she would suggest to be the pedagogical qualities of many years of place research in close collaboration with Australian aboriginal people. Her formulation of key elements of a pedagogy of place was underpinned by empirical qualitative studies and her own learning from this collaboration. She had learned to think *through place* in her own formulation. Key elements in the conceptual framework for her suggestion of place pedagogy are "our relationship to place is constituted in stories (and other representations); the body is at the center of our experience of place; and place is a contact zone of cultural contact," (Somerville, 2010, p. 335). These elements must all be simultaneously present and do not have any hierarchical ordering. Ritchie (2012) holds that research describes an ongoing "un-storying" of the world, where especially indigenous stories interwoven with nature and culture are marginalized. Both colonization by force/war and modernist Western globalized culture have served to marginalize Maori traditional knowledge and histories, even if valued by local teachers and community members (Ritchie, 2012). Duhn (2012) argues that to work with an understanding of place as assemblage requires a pedagogy that is cognizant of the assembled nature of *all* matter that makes up a *place*. She utilizes Jane Bennett's notion of vibrant matter in this exploration, along with Deleuze and Guattari's (1987/2004) emphasis on intensities, becomings, and assemblages (Bennett, 2010).

## *Researching place pedagogies—methodological considerations*

As we worked with place-based ideas, we asked the question: what kind of methods are reasonable in doing research where place as assemblage, indigenous perspectives, and pedagogy are bundled together? Turning to Law (2004), who argues that the research methods of present social sciences are not beneficial/adequate/fit to handle complexities and "mess," we considered the networks of Sámi. The places we study, the networks where social and material elements are assembled, for example, were/are continuously changing. This means that methods should be able to meet the ambiguity and multiplicities that characterize both places and pedagogies. According to Law (2004), this is something we as researchers have to work on and develop. In our studies, we use traditional methodological approaches, fieldwork, observations, stories, and interviews. Law (2004), who refers to method in an extended manner, speaks of "method assemblage" in his book *After Method*. The term "assemblage" is the English translation of Deleuze and Guattari's "agencer" which has a broader meaning than assemblage has in English (Law, 2004). Law explains, with reference to Watson-Verran and Turnbull, how assemblage in its broader sense can include many elements. Assemblage is like an episteme with technologies added but that connotes the ad hoc contingency of a collage in its capacity to embrace a wide variety of incompatible components. It also has the virtue of connoting active and evolving practices rather than a passive and static structure,

(Watson-Verran and Turnbull, 1995 in Law, 2004, p. 41)

As we studied with Sámi, we were in need of both the *capacity* and the *virtue* mentioned in Law's understanding of assemblage. Law (2004) constructs *method* assemblage, which is distinguished from *assemblage* in a so-called provisional definition, as the enactment of "a bundle of ramifying relations that generates representations in-here and represented realities out-there" (p. 14). Drawing further upon Deleuze and Guattari (1987/2004), "[T]ools only exist in relation to the interminglings they make possible or that make them possible" (p. 99), we read this to be in line with Law (2004) and his arguments that methods are performative and produce the realities that are researched. In other words, "Realities are produced along with the statements that report them," (Law, 2004, p. 38). Law (2004), who draws again on Deleuze/Guattari as well as Latour/Woolgar, is also concerned with the production *within* the methods in use, the multiplicity, fluidity, and relationship between epistemology and ontology.

Drawing upon Law, Deleuze and Guattari, Ritchie, Duhn, and Somerville, we forged connections of Sámpi place like "a bundle of ramifying relations" (Law, 2004, p. 13). Laila Aleksandersen Nutti, who is from the northern Sámi area, created interviews which were carried out with Sámi elders. Bushra Fatima Syed and Aslaug Andreassen Becher, from Oslo, created interviews with the pedagogical leader and a resource person for the Sámi kindergarten. The leader was interviewed twice to gather more detailed information. Additionally, Syed created further fieldwork, following the kindergarten children, parents, and the leader on a trip up north to Sápmi. As researchers, two of us have an "outsider" position (Bushra and Aslaug) and one an "insider" position (Laila) relative to the indigenous knowledge and perspectives constituted by research with Sámi. We are all "Norwegians," but only Laila has Sámi family ancestors: she has lived her life in Sápmi, speaks the Sámi language, knows Sámi crafts. What we see, how we understand what we see, and in what ways we connect theories to what we see are influenced by our positions. However, our educational and professional backgrounds, and the fact that we work in academic institutions, also lead us to reflect upon what we see in equal and sometimes different ways. There are ethical considerations to undertake when understanding indigenous perspectives, because such groups have been historically marginalized. We were careful to be authentic in the production of text, and the use of pictures was with consent and mutual understanding with our informants. We respect the persons who have shared their experiences, reflections, relations, and places with us. This is especially important when it comes to elders and research in indigenous perspectives (Smith, 2012).

The areas, places, and institutional surroundings of Sápmi are different, but also share similarities. The climate and the urban/rural communities differ, but particular signs of Sámi connection will be similar across place. Today there are Sámi early childhood institutions in rural communities in Sápmi, as well as some few in urban areas, like the one in Oslo. Traditional practical work is often part of the content in these institutions even though the contexts and access to geographic places are diverse. Images in Figures 4.1 and 4.2 show elements in the Sámi kindergarten which invite children to play with and learn traditional Sámi practices like gathering for smoking reindeer meat, fishing and boating, and sheltering in the Lávvu.

**FIGURE 4.1** Gamma in the kindergarten in Oslo city
*Source*: Aslaug Andreassen Becher

**FIGURE 4.2** Children playing in the Riverboat in the kindergarten in Oslo city
*Source*: Aslaug Andreassen Becher

We had five informants: two from Oslo and three from the northern areas of Norway. The informants were all strategically requested to participate in this research because of their experiences, traditional knowledge of crafts, and indigenous philosophies/thinking. The informants have worked in Sámi early childhood institutions for many years and some were involved as resource persons, both for educators and children during traditional activities such as fishing or other traditional crafts. The elders are resource persons because of their special language and cultural knowledge. The purpose of this arrangement (the elder role of these particular knowledge holders) is to protect and secure elements of the historical knowledge and indigenous skills as they are passed on to new generations. There are special funding programs promoted by the Sámi Parliament and Sámi organization that secure the present arrangement between elders and students. Additionally, some informants are grandparents in their own or in extended families which are still common in areas. Some are also "grandparents" for the children in the early childhood institutions. These kinship networks have, in different ways, transformed their Sámi knowledges and work to fit within institutional settings. Focusing on traditional work, philosophies, and activities both out on the land and inside the institutional houses/areas together with the children, several issues were highlighted with ideas and reflections expressed about Sámi pedagogic settings. The role of Sámi institutions was reflected upon and discussed during the interviews as well.

Informants received an outline of our questions/themes prior to their interview. However, the interview guide was not strictly followed, and as we became engaged in the informants' narratives, we allowed ourselves to ask additional questions. Since our interests were related to multiple issues of materiality, crafts, ages, practices, culture(s), traditions, and nature, there were many variables that could be traced in the research material. For the conversational interviews, some primary topics were discussed, but the informants were free to talk about other issues as well, and they moved freely between issues that captured their interest during the interviews.

## In what ways are place-based pedagogies traced in the kindergartens?

Informants in the early childhood institutions tell us that place is concrete and connected to land and territories. Place consists of nature—human—and culture. Survival and sustainable living in, with, and by nature has been part of the Sámi traditional form of life for thousands of years. Accordingly, what nature affords in various places creates different knowledges, clothes, food/dishes, crafts, arts, and vocabulary. Place as location is significant, but at the same time undefined and fluid. Place is also embraced in spirituality as "sense of place" but not static in its inherency. This spirituality is expressed, for instance, when the elders do thanksgiving to a "Sámi place" in the forest.

### "Sámi places for survival" in Northern Sámi regions becoming children's places for play and learning

One informant (Inf. 3) from the Northern Sámi region told us that elders would usually look for places as close as possible to the early childhood institutions. Issues concerning families' use, interdependence, and connection to the place needed to be taken into consideration. In the Sámi tradition and language, there is no concept for "the right to property," one of our informants (Inf. 1) maintained. The elders shed light on the kind of ownership that accrues from the use of places, through the generations. One of the elders reported that it was fine for him to help the educators because he would be with the children in the places that he regularly visited: "It felt good and easy to join the work and the trip; when the educators decided to go to a lake I usually would go too. It was very familiar to me."

One of the educators (Inf. 4) discussed the advantage of the early childhood education institution having one particular place to go net-fishing, for example. Elders and children could communicate and collaborate with people living close by; they could leave the boat and build their own connections to the place by choosing their own spots for traditional activities. In the informant's words, "it kind of got to be our own place that we found and chose the spot for the fires place and for the lávvu."

The Lávvu becomes for the youngest children a place to go where the bun is prepared for smoking the meat. The children are following and participating in the whole process by setting up the làvvu, arranging tennvu, taking care of the fire, hanging the meat in threads and making sure everything goes right when the reindeer meat is being smoked (see Figures 4.3 and 4.4).

## Lávvu play with kindergarten children

Both the educators and elders talked about the value of building long-term relationships with these places, where they returned several times a year. The possibility to connect with these places through regular visits created familiarity. "Both children and we adults got to know the area and the places and the children kept on exploring and look for new things to do when they got to come back several times."

**FIGURE 4.3** Two children looking into the Lávvu where the bun is prepared for smoking reindeer meat

*Source*: Hellander

**FIGURE 4.4** View from within the Lávvu
*Source*: Hellander

In researching Sámi kindergarten practices, we find that place-related knowledges are inter-woven with the Sámi language. Our informants (Infs. 1 and 2) explain how one single word in the Sámi language can describe a decrease of a slope, the temperature in the ground, or the health of a reindeer. These concepts are tied up with life and traditional activities in connection with nature. In a way, the values, knowledge of crafts (ways of living, sustaining), and language are interwoven with the Sámi sense of place. The words express the values and the necessary actions to maintain the traditional knowledges and skills and to pass them on.

## Sápmi consists of Sámi places

> To be a Sámi is about identity, and you can easily identify with both Kautokaino, Karasjok, and Tana [small towns in the county of Finnmark, in the middle of Sápmi]. But it is also about what is in your heart!
>
> *(Inf. 2)*

Here the informant references the world-renowned artist Nils Walkepää, who said that place is both geography and what you keep in your heart. Informant 1 was asked what actually constitutes "Sáminess"? The informant responded that "real old Sámi people" argue that Sámis originally were Norwegians who moved to the place where the Sámi lan-guage was spoken; they learned the language, the crafts, and the traditions of the place and then became Sámi. However, after the Norwegian colonizing processes during the 18th, 19th, and 20th centuries, Sámi people felt the need to protect their origins in more distinct ways, which for many implied that you did not become a Sámi simply by living as one. This can still be a contested issue among Sámi elders.

Places in Sápmi where Sámi traditions are practiced and expressed have an impact on Sáminess, according to our informants. To interweave traditional Sámi "sense of place" into a kindergarten located in the middle of the capital city, the 6-year-olds are taken on an annual trip to the villages and places in Sápmi. Bushra (Syed) was allowed to accompany the children, the staff, and parents on a trip to Sápmi. They visited places where they could experience activities characterized as "indigenous," and which had the purpose of giving the children firsthand knowledge about culture and crafts in surroundings that differed from Oslo.

Here the children had the opportunity to visit places where their relatives, grandparents, aunties, and uncles lived. Figure 4.5 shows a visit at grandmother Ingá's home. These places had the potential to connect them to their cultural heritage. Riding horses and visiting grandparents were activities s that might strengthen the connection between Sáminess and place, both as location and in the heart.

## Places that become "Sámi" in the forest surrounding Oslo city

*"Typical Sámi thinking is that there is a close connection between us and place. And that good experiences strengthen this connection. In this way, Ulsrudvann [in Oslo] can become a Sámi place"* (Inf. 2).

Elders and educators spoke of the *becoming* made possible with repeated visits to Sámi places. As one explained,

> with concepts that connect to nature, it is possible to be Sámi and have a Sámi relation. [And] to "make a place a Sámi place" … We have our own place in the forest … a place that we come back to. Where we sometimes set up the lávvu, where we do ice fishing, where we fry meat from the reindeer. A place in the woods that the children recognize.
>
> *(Inf. 2)*

**FIGURE 4.5**  Visiting grandmother Ingá's home. Having a nice time in front of the barn
*Source*: Bushra Fatima Syed

The kindergarteners travel by bus to their place in the forest. A "grandfather" and "grandmother" accompany them on these tours. The grandparents/elders have the traditional Sámi knowledge of how to act in nature: They know the crafts, the tales, the rituals, and how to "joik" the special songs to nature, places, animals, or life themes. They also show the children how to use the drum and the knife, and how to make food in traditional ways. "The elders are our libraries. They bear knowledges that you do not find in books," Informant 2 claims. "All this the children experience and are encouraged to learn from. When they leave the place, grandmother 'thanks' the place and makes sure that there are not any traces left from their activities."

> Grandmother is very concerned about our leaving the place—so we shall do thanksgiving that we were allowed to come back to this place. We shall thank nature for branches to the campfire. She wants the children to do thanksgiving. She will say, "Thank you for allowing us to come here. We will soon be back."

Our informant described the work of leaving the forest. "The fireplace in the forest must be clean, not soiled or littered. [We must] have respect for nature and the gifts nature gives us" (Inf. 2). Informant 1 explains that the traditional relationship to nature requires that there shall be no traces of you left behind. "You shall also appreciate all the participants in the landscape," he says. In the traditional understandings, "participants" are living souls and part of a total community. This shows that spirituality is interwoven with the concept of place.

As our research continued, we had conversations on how places can become a *place*. We asked if there were any special circumstances that needed to exist for a place to become *their place*. Our informants told us that several places can *become a Sámi place*. Places that have materialities that can be conceptualized in the language have such potential. For example, the kindergarten "grandfather" found a new place last autumn with the traditional *senna* grass in another location around Oslo, which also is becoming a Sámi place through the actions of the elders and children. Senna grass has been used for hundreds of years to keep feet warm during wintertime. The grass will be cut and manipulated in a particular way, so it becomes soft and pliable for use as a sole in reindeer-hide shoes, called *skaller*. It has been proven that skaller filled with senna grass will keep feet warm even if the temperature is minus 20 degrees Celsius. Children learn how to cut and process the grass from these senna grass field trips (see Figure 4.6, 4.7). They also learn that senna grass is sustainable and kind to the environment. "Nature will take it back without any traces of littering."

The kindergartens are working with "storying," which means that children and participants relate to the community they form on trips. The trips are prepared in advance by recalling, in different ways, what happened in past years' experience. The parents are positive and grateful that the kindergarten is providing their children with traditional knowledge and experience, and the pedagogical leader tells us "Grandfather" and "Grandmother" are important storytellers when the kindergarteners go to "places" around Oslo city. Simultaneously, children experience the place as filled with material elements, such as the lake/water, fish, fireplace, reindeer meat to be grilled and grass to be cut and processed. This materiality invites children to act and experience. Their "guided experiences" of *place* are connected to indigenous forms of life and knowledge. Unlike the traditional kindergartens, these experiences s allow the children to construct a meaning of love for the materiality of the earth and the place of Sámi.

**FIGURE 4.6** Senna grass being moved and hit for craftwork
*Source*: Laila Aleksandersen Nutti

**FIGURE 4.7** Senna grass being processed before it is used
*Source*: Laila Aleksandersen Nutti

### Places help connecting traditional crafts to the position of elders and their epistemological and ontological knowledges

The places where Sámifying can happen is of great importance and a condition to vitalize the traditional knowledge. Learning the practical crafts is important, but even more important is the discourse of place that comes with the craftwork: meaning is everywhere and can be captured by incorporated signs and signals.

One informant said,

> [Outdoors on a nature trip] the children become so good at feeling and knowing by themselves if they are feeling too hot or too cold … and then they realize that it was not a good idea to leave their mittens and hats all over the place.

The children are allowed to experience how clothes are necessary to master the cold climate in these surroundings. Being out in the natural world, working with something over time, models the realities of real work in a real way, and retreating indoors is often not an option.

### Conclusion: analyzing assemblages and place-sensitive pedagogy

*Place* in context is concrete and connected to land and territories–Sápmi areas. Place consists of the intersection of nature, human, and culture. Survival and sustainable living in close connection to nature has been a Sámi tradition for thousands of years. Accordingly, what nature affords in a place, combined with human activity, has formed a variety of Sámi knowledges, clothes, food/dishes, crafts, arts, and vocabulary words. Traces of an understanding of *place as location* is significant. Place in Sámi contexts is also filled with spirituality. "Sense of place" is part of the discourse, but this is not fixed. Informants expanded upon how Sámi thinking reflects cycles and the movements of seasons, creatures, and people. Our informants expressed that such cyclic thinking can mean openings to new beginnings. "Nothing will be forever the same—ever" (Inf. 2).

*Place* is accordingly a fluent concept. It can be created and recreated in several locations and thus has the unfolding character that assemblages have. Interviews with educators, elders, and resource persons in Sámi early childhood institutions in both parts of the country reveal an argument that *place* can be "profoundly pedagogical." *Places* as centers of experience "[can] teach us about how the world works, and how our lives fit into the spaces we occupy," (Gruenewald, 2003, p. 647). By following a poststructural and posthuman perspective, like Somerville's, we see that humans are working *on* and *with* place and thus making the "world as place" change. In this fashion, places are *becoming* in relation to people and both living and non-living creatures. Places have the embedded possibilities of the unfolding of differences as in Massey's or Duhn's perspectives. Simultaneously, there is a sort of "distinctness" traced out by our informants that may point to some stability in material elements and affiliation between human and place as it is embedded in the general aspect of *place as location* for the Sámpi.

Because of our findings, we still ponder whether indigenous knowledges about humans and non-humans can produce meaning in present early childhood settings. Kincheloe and Steinberg (2008) argue that indigenous knowledge can, in itself, provide a critique of Western ways of knowing and being. This would be in line with Linda Tuhuwai Smith's (2012) call for new decolonizing methodologies and methods for academic work as a contribution to research and understandings acknowledging indigenous peoples and practices/traditions. We recognize that most of the traditions are no longer essential for survival in present Western ways of life, and that most children do not experience traditional skills and tasks in their homes, such as the ones we discussed. However, it is important to explore the value of such content as part of the pedagogic process. You do not need to fish for food; you can buy the fish. And you don't need to fill your boots with grass, because you can buy warm socks. However, our informants seemed to feel that through these kinds of work—often taking place outdoors, in interaction with nature and the elements—it was easier to pass on knowledge, language, and values of Sámpi in traditional ways. Elders and kindergarten teachers believe this is a way of strengthening the children and their Sámi identity. The traditions are unfolding in a place that becomes the "zone of cultural contact" (Somerville, 2010). The body is at the center of the learning because all the skills rely upon the use of the body. Language and stories accompany the activities that represent, expand upon, and ultimately strengthen the children's experience of identity.

## Openings for the future—writing and doing "place" as part of the Norwegian curriculum

The Norwegian early childhood curriculum is not yet following the structure of detailed "learning outcomes," as is increasingly common in every educational field in Europe, the U.S., and Australia. Our informants were asked whether they would prefer a more elaborated educational policy on *place* in the curriculum. However, they regarded the non-standardized curriculum of today as adequate, because it allows for a freedom in learning and practicing the epistemologies and languages of Sámi. Our informants were not concerned with the danger of underpinning stereotyped perceptions of Sámi traditional practices. They argued, instead, that it is necessary to show some of the distinctness of Sámi culture, places, and expressions, even while those expressions of craft, story, and survival are no longer based so much in present-day necessity as perhaps in history. Here, other educators and researchers might disagree or point to the danger of producing essentialized perceptions.

Kincheloe and Steinberg (2008) write about the transformational power of indigenous knowledge, the way indigeneity can be used, the benefits, and how to ensure that the interests of indigenous people are served. We wonder whether the Sámi knowledges described herein might have the potential to form a resistance toward the pressures of producing detailed learning outcomes, regulated childhoods, and constrained images of childhood such as are now common around the world. As critical early childhood researchers, educators, and kindergarten teachers in several parts of the world, we imagine others would probably embrace a move such as "places becoming Sámi," because such a move would challenge and disrupt standardizing and dominating discourses of childhood. The indigenous elders and experienced "resource persons," like the ones we studied, should be given an important and respected role in overcoming kindergarten standardization.

We believe that through the words in curriculum mandates across the world related to epistemologies and languages of *place*, we might develop a more powerful resistance toward mainstream national curricula that are *not* inclusive of place-specific knowledges. Indeed, examples of the Sámi would follow and grow just as the example of Te Whāriki. The call for renarrativization in kindergarten pedagogy would also relate to *place as vibrant matter*, becoming fluid and diverse (Bennett, 2010; Duhn, 2012). When *place* is analyzed in the perspective of assemblages with humans and their "multiple others" elements, networks of relations become possible. In stories of *place*, the anthropocentric stories of humans as the center of activity become less dominant. An emphasis on the forces and vibrancy of materiality in teaching young children can be included in traditions and the narratives of *place*. Within this framework, we can acknowledge how the world acts upon the Sámi knowledges, while simultaneously the Sámi sense of place acts upon us in the world.

# References

Agnew, J. A. (1987). *Place and politics: The geographical mediation of state and society.* London: Allen & Unwin.

Aikio, A. (2010). *Olmmoshan gal birge: ássit mat ovddidit birgema.* Kárásjohka: CálliidLágadus.

Balto, A. (1997). *Sámi mánáidbajásgeassin nuppástuvvá.* Oslo: Ad Notam Gyldendal.

Balto, A. (2008). *Sámi oahpaheaddjit sirdet árbevirlas kultuvrra boahttevas buolvvaide: Dekoloniserema aksuvdnadutkamus Ruota beale Sámis.* Guovdageaidnu: Sámi allaskuvla.

Becher, A. A. (2006). *Flerstemmig Mangfold: Samarbeid med minoritetsforeldre.* Bergen: Fagbokforlaget.

Bennett, J. (2010). *Vibrant matter: A political ecology of things.* Durham, NC: Duke University Press.

Bergstrøm, G. G. (2001). Tradisjonell kunnskap og samisk modernitet. En studie av vilkår for tiegnelse av tradisjonell kunnskap i en moderne samisk samfunnskotekst. Hovedoppgave i samfunnsvitenskap, Universitetet i Tromsø.

Castree, N. (2009). Place: Connections and boundaries in an interdependent world. In S. L. Holloway, S. P. Rice, & G. Valentine (Eds.), *Key Concepts in Geography* (2nd ed., pp. 165–185). London: Sage.

Cresswell, T. (2004). *Place: A short introduction.* Malden, MA, Oxford, UK, Victoria, Australia: Blackwell Publishing.

Deleuze, G., & Guattari, F. (1987/2004). *A thousand plateaus: Capitalism and schizophrenia* (B. Massumi, Trans. 2004 ed.) London: Continuum. (First edition *Mille plateaux*, vol. 2 of *Capitalisme et schozophrénie*, 1980).

Duhn, I. (2012). Places for pedagogies, pedagogies for places. *Contemporary Issues in Early Childhood, 13*(2), 99–107.

Førde, A., Kramvig, B., Berg, N. G., & Dale, B. (2013). Introduction: Methodological challenges in analyzing place. In A. Førde, B. Kramvig, N.G. Berg, & B. Dale (Eds.), *Finding place: Methodological perspectives in analysis of place* (pp. 9–22). Tronheim: Akademika Forlag.

Greenwood, M., & de Leeuw, S. (2007). Teaching from the land: Indigenous people, our health, our land, and our children. In A. Pence, C. Rodriguez de France, M. Greenwood, & V. Pacini- Ketchabaw (Eds.), *Indigenous Approaches to Early Childhood Care and Education* (Vol. 30). Alberta: Canadian Journal of Native Education.

Gruenewald, D. A. (2003). Foundations of place: A multidisciplinary framework for place-conscious education. *American Educational Research Journal, 40*(3), 619–654.

Guttorm, G. (2011). Árbediehtu (Sami traditional knowledge) as a concept and in practice. In I.J. Porsanger, & G. Guttorm (Eds.), *Working with traditional knowledge: Communities, institutions, information systems, law and ethics* (Vol. 1/2011, s. p. 17). Guovdageaidnu: Sámi allaskuvla/Sámi University College: Dieđut.

Hirvonen, V. (2009). Juoiggus uksan sámi njálmmálas girjjálasvuhtii: bálggis gillii, identitehtii ja iesárvui. *Sámi dieđalaš áigečála, 1*(2), 90–105.

Hirvonen, V., & Balto, A. M. (2008). Samisk selvbestemmelse innen utdanningssektoren. *Gáldu čálá, 2*, 21.

Holloway, L. & Hubbard, P. (2001). *People and place: The extraordinary geographies of everyday life.* Harlow, UK: Prentice Hall.

Kincheloe, J. L., & Steinberg, S. R. (2008). Complexities, dangers, and profound benefits. In K. Denzin, Y. S. Lincoln, & L. T. Smith (Eds.), *Critical and Indigenous Methodologies* (pp. 135–156). Los Angeles, CA, London, New Delhi, Singapore: Sage.

Kuokkanen, R. (2000). Towards an "indigenous paradigm" from a Sámi perspective. *The Canadian Journal of Native Studies, 20*(2), 411–436.

Kuokkanen, R. (2007). *Reshaping the university: Responsibility, indigenous epistemes, and the logic of the gift.* Vancouver: UBC Press.

Law, J. (2004). *After method: Mess in social science research.* London, New York: Routledge.

Massey, D. (1994). *Space, place and gender.* Cambridge, UK: Polity Press.

Massey, D. (2005). *For space.* London: Sage.

May, T. (2005). *Gilles Deleuze: An introduction.* New York: Cambridge University Press.

Melhus, C. (2012). Hytta i skogen. Et sted som bli til (The cottage in the woods a place which emerge). In Krogstad et al. (Eds.) *Rom for barnehage. Flerfaglige perspektiver på barn og rom. Room for Kindergarten. Multiple Perspectives on Children and Room* (pp. 193–212). Bergen: Fagbokforlaget.

Nergård, J.-I. (2005). *Den levende erfaring: En studie i samisk kunnskapstradisjon.* Oslo: Cappelen akademiske forlag.

Nutti, L. A. (2011a). Juoiggadan – Jeg holder på så smått med å joike. Med urfolksblikk og re-tenking rundt barnehagen som joikearena Master, Høgskolen i Oslo og Akershus.

Nutti, L. A. (2014). Samiske elever – en del av skolens mangfold. In I.B.B. Moen (Ed.), *Kulturelt og etnisk elevmangfold i skolen* (pp. 155–166). Bergen: Fagbokforlaget.

Nutti, Y. J. (2011b). Ripsteg mot spetskunskap i samisk matematik : lärares perspektiv på transformeringsaktiviteter i samisk förskola och sameskola. Ph.D., Luleå tekniska universitet, Luleå.

Nutti, Y. J., & Joks, S. (2018). En barnehages praktiske engasjement former stedets karakter. In A. Myrstad, T. Sverdrup, & M. B. Helgesen (Eds.), *Barn skaper sted – sted skaper barn* (pp. 189–201). Bergen: Fagbokforlaget.

Rameka, L. (2012). Te Whatu Kakahu: Assessment in Kaupapa maori early childhood practice. Doctor of Philosphy, University of Waikato.

Rameka, L. (2016). Kia Whakatomuri te haere whakamua: I walk backwards into the future with my eyes fixed on my past. *Contemporary Issues in Early Childhood, 17*(4), 387–398.

Ritchie, J. (2012). Early childhood education as a site of ecocentric counter-colonial endeavour in Aoteraroa New Zealand. *Contemporary Issues in Early Childhood, 13*(2), 86–98.

Rowan, M. C. (2017). Thinking with Nunangat in proposing pedagogies for/with Inuit early childhood education. Doctor of Philosophy, The University of New Brunswick.

Smith, G. H. (2004). Mai i te Maramatanga, ki te Putanga Mai o te Tahuritanga: From conscientization to transformation. *Educational Perspectives,* 46–52.

Smith, L. T. (2012). *Decolonizing methodologies: Research and indigenous peoples* (2nd ed.). London: Zed Books.

Somerville, M. (2007). Postmodern emergence. *International Journal of Qualitative Studies in Education, 20*(2), 225–243.

Somerville, M. (2010). A place pedagogy for "global contemporaneity". *Journal of Educational Philosophy and Theory, 42*(3), 326–344.

Storjord, M. H. (2008). Barnehagebarns liv i en samisk kontekst: en arena for kulturell meningsskaping. Ph.D., Universitetet i Tromsø, Tromsø.

# 5

# TURKANA INDIGENOUS KNOWLEDGE AS NARRATED BY TURKANA ELDERS

Implications for early childhood curriculum in pastoralist communities in Kenya

*John T. Ng'asike*

## Turkana traditional cultural practices and indigenous epistemologies

During this ethnographic study, I interviewed Turkana elders (women and men) as well as observed children pastoralist livestock herding activities in the natural settings of their everyday life styles (Maanen,1988; Malinowski, 1922; Ng'asike, 2010). The discussions with the elders focused on their indigenous traditional knowledge of hydrology, weather and climate, livestock, plants, tools, traditional medicine, and socialization of children (Ng'asike, 2010).

The *Turkana Calendar* emerged as key determinant of Turkana people's everyday life pastoralist activities. According to the elders, Turkana people believe that the moon rises and "dies" 28 eight days later on the twenty-eighth day (of each month) and rises again on the thirtieth day. The twenty-eighth day is also the end of the Turkana lunar calendar. The death of the moon causes complete darkness for one day. On the second day the moon rises but only blind people can see and everyone will see the moon on the third day (31st or 1st). The Turkana calendar will start to operate on the third day when the moon is visible to everyone in the community. Turkana people do not count "the dark" day when the moon is not sighted. The difference is the Western calendar includes the dark days when the moon is not in sight. Each Turkana year has six months consisting of dry and wet seasons respectively. The elders argued that after interacting with the Western calendar, the community accepted modifying their calendar and merged their six month seasons into one full year calendar of twelve months to match with the Western calendar, even though their calendar year starts in March and ends in February. The wet season starts in March and lasts through August, followed by a dry season that begins in September and ends in February. In the Turkana language, the wet season is called *Akiporo* and the dry seasons is referred to as *Akamu*. The names of the months in the Turkana language describe the weather conditions as reflected by changes in the ecosystem. Some months' names signify the activities the people carry out during those particular weather conditions. For example, there are some months the community devotes to rituals and celebrations, while in others people engage in coping activities as the weather becomes drier and food becomes scarce. The names of the months of the Turkana Calendar are: *Lomaruk* (March), *Titima (*April), *El-el* (May), *Lochoto* (June), *Losuban* (July), *Lotiak* (August), *Lolong'u* (September), *Lopo* (October), *Lorara* (November), *Lomuk* (December), *Lokwang* (January), *Lodunge* (February).

The Turkana study the universe, star patterns, moon, and the Sun colors to predict rain, drought, or a calamity (e.g. death of a community leader). Eclipses are regarded as the death of the moon or of the sun. When eclipses of the moon and the sun are sighted, every household in the entire community plays the drum (played by women assisted by children in all the villages). The sounds of drums travel across the entire community, and the different villages respond in unison to the drum beats. The purpose of drumming is to resurrect the moon or the sun from death. The drum beating stops and people return to their normal socioeconomic activities as soon as the eclipse clears.

One elder made a comment, "today the world is upside down," indicating that in this modern era they are surviving merely through luck as they are entirely dependent on the government, which in many instances is not close to them. The elders argue that they are unable to survive on their own as they are no longer able to predict and monitor their own weather patterns using their traditional methods. These observations have been noted in many indigenous people in Africa where communities are no longer able to use the traditional methods they are familiar with in making adjustments on climate change (Eromose & Danny, 2017; Ng'asike & Swadener, 2015). Based on their predictions Turkana people are able to manage the grazing patterns of their livestock (e.g. move the herds to a different location when necessary) and make offerings to keep the community shielded from calamities (man-made or natural).

The Turkana use the stars and the planets at night to locate the directions of their destinations. The stars at night are also used to approximate the hours of the night. Turkana people are travelers and therefore knowing the hours of the night and the direction of their destination is very critical in figuring out their routes. Turkana experts read the intestines of both domestic and wild animals to confirm the predictions of weather changes that were predicted by observing the sun or the moon. Turkana people predict natural phenomena such as rain, weather conditions, calamities, visitors, drought, enemies, and killer diseases by reading the intestines of goats, sheep, cows, and wild animals such as rabbits and gazelles—camel and donkey intestines are not used for making predictions. The elders describe the intestines as an animal organ that houses their worldview. When a family slaughters a goat for consumption, the intestines are removed before they are cooked and spread on a stone or on a flat surface. The elders study the intestines very keenly in the same way a map is read by the experts—the difference is that the intestines represent a dynamic worldview of the people and show events in the community as they unfold continuously. In the intestines, they can read the settings of the community, the villages, and homesteads. The intestines will show in detail every aspect of the homestead, for example, the settings of the compound, number of co-wives, arrangement of their houses, number of children, and relatives. The intestines will provide information, such as expected number of visitors, fortunes, and misfortunes. The intestines will tell the origin of the visitors and the direction from which they will enter the village or homestead including an impending attack by enemies to the community.

The men are usually the intestines readers, even though women are capable of doing the same with accuracy. Women read the intestines when they are in charge of the family or when they are permitted by their husbands. The community trusts the information the elders communicate from the results of reading the intestines. In the present contemporary lifestyle where modernity is the way of life, elders have lost the ability to use their cultural practices to the extent that the use of intestines for prediction is becoming forgotten. In recent times, global warming has led to unpredictability in the climate in Turkana. Droughts are becoming very common; the elders and the communities are becoming very vulnerable. The elders believe that modern life has interfered with the way they strategize traditionally to respond to issues they predict. For example, even if they were to apply their predictions to organizing themselves, they would no longer depend on their own experts for solutions. They have to report everything to the chief or to other local political leaders who usually do not trust traditional ways of predicting Mother Nature. In addition, modern ways of leadership contradict traditional modes of administration.

The children receive training for all the skills elders practice by observing adults. The children learn how to read the intestines and make predictions as soon as they start herding livestock, which is usually during the preschool age. For example, by toddlerhood a child is watching and holding a stock such as the kid of a goat. The mother and the father will encourage a child very early on to touch a goat, a sheep, or the kid of a goat. By the time the child is two and half years she/he is already following older children or the parents when carrying out livestock activities, such as maintaining the health of the herd, finding water and grass. Parents, elders, and clan members have the responsibility to train young people to ensure that the skill is maintained by the clan. Theories on observational learning arguing on peripheral observers (Lave & Wenger, 1991) support the modes of learning demonstrated by pastoralist children.

Women and men have equal responsibility in training children in all the cultural skills. The training of women is the most efficient and thorough. A child's interest determines the kind of training that he/she will acquire from the elders. Children follow the adults and observe and internalize every activity they do, including how to slaughter an animal, skinning, roasting, reading the intestines, and others. The children will learn the rituals and spiritual activities related to every event the adults perform. Boys and girls learn all the cultural skills needed for competency. All children are recognized according to their efforts and the competency at which they learn the skills they are taught. Although every young person has a general knowledge of the concepts and skills of his/her culture, it is not mandatory that children learn every skill to perfection. Young people perfect only the skills they develop interest in from early childhood.

## Livestock herding

Herding of livestock is a specialization children develop interest in and associate themselves with from childhood, with a particular herd or species of live livestock. Turkana people keep five species of livestock: goats, sheep, cattle, camels, and donkeys (McCabe, 2004). Donkeys are the beast of burden, although in some families they are slaughtered for meat. Donkey milk is given to young children for immunity against respiratory diseases. Goat, sheep, camel, and cows' milk is consumed by both adults and children on a daily basis. Most rituals are performed by slaughtering a goat or sheep. These species of livestock are also sold for money, which is used to buy the family other food items (sugar, corn, beans etc.) and tobacco. In addition, the livestock are slaughtered when rituals and ceremonies are performed for members of a family and community offerings. As early as in preschool years the children are already capable of taking the herds to the pastures and water points.

Treatment of livestock is by the use of herbs which they crush to squeeze out the extract for the animal to drink or to smear on the wound. The plant's herbs may be boiled and the extract given to the animal. Tsetse fly, ticks, and Trypanosomosis are common parasites that cause livestock diseases. Livestock that die as a result of disease attack are eaten. This fact was emphasized by the elders and it is an issue to follow up with the elders in the future. The pastoralists believe that livestock diseases are increasing in the modern era. They argued that modern lifestyles have contributed to an increase in diseases because they know they have the medicine for treating them. This remark from a Turkana elder is important to note:

**FIGURE 5.1** Youth in pastoral roles in Turkana communities
*Source*: John T. Ng'asike

Even rain has become scarce, because white people know that they have relief food to give us. There used to be Locust which used to destroy all the grass in the community. Livestock suffer as a result of the locust; however, rain comes after the locust resulting to large pastures. Today you cannot see the Locust. It was destroyed by the airplane which powered a yellow drug on the land to kill all the Locusts. After that we no longer see the Locust. This has also lead to frequent drought which has drastically reduced the pasture which used to bring the Locust to our land.

*(Ng'asike, 2010, p.114)*

## Herding skills training

Turkana people get attached to the livestock physically and psychologically to the extent that every behavior and movement of the stock is predictable. For example, children, as well as adults, can recognize their stock by studying the prints and marks of the hooves left on the ground by the animals. The Turkana will tell by observing the print marks of the hooves of the animal on the ground whether this stock belongs to their herd or is straying stock from a different herd. Turkana herdsmen study livestock hoof prints in almost equal accuracy similar to the scientist study of figure prints. An elder turned my shoe upside down and explained,

[D]o you see this shoe, do you see it has prints, the camel's hoof is the same. Each camel has prints unique to itself. By studying their hooves, we can tell our camels when it gets lost or when it is stolen by another herdsman by tracking the hooves' prints.

*(Ng'asike, 2010, p.115)*

By studying the hoof marks on the ground, they will tell that this animal is from the lowland or from the mountains/highland. This is a skill the children have to master as they interact with the livestock from childhood. Turkana herdsmen use these unique marks to trace specific livestock of their herd when they are lost or stolen.

Turkana people read the footprints of people in the same way they track livestock. Individual herdsmen master the shoe marks of their relatives, friends, and neighbors. Turkana herdsmen use the same pair of shoes and as a result it makes it easy to master the footmarks of individuals. Studying people's footprints is important, because livestock stray with individuals. Theft of livestock among the pastoralists is a common cultural practice, resulting in frequent conflicts and livestock rustling among neighboring pastoralist communities. Tracking of people will help the herdsmen to know the presence of strangers in the community as they are likely to be a threat to their livestock.

Turkana pastoralists study the anatomy of their herds. They can tell the body structure of the livestock, the length of the bones, and general weight of a particular stock. They have such complex knowledge of livestock including fertility and gestation period, dietary intake, and species of forage intake capacity. Elders described a strange situation in which certain individuals adapt to the livestock to the extent that they can tell from testing the meat that the meat belongs to a stock from his herd. These individuals will examine the bones of the dead stock in addition to the taste of its meat to confirm evidence that indeed the stock was slaughtered from his herd without his/her knowledge. Livestock are also branded and a family will have a general brand that identifies the livestock species of their herd.

Instead of keeping number records of livestock, the Turkana classify the herds according to cohorts. Livestock species reproduce in seasons and the herdsmen keep track of the young ones born every season. They classify them into a category or cohort and observe, learn, and master their characteristics as they develop to maturity. For instance, in each cohort, the herdsmen know the males and the females including the genotype and phenotype characteristics of the livestock species. When this cohort reproduces, another cohort emerges which is described as the offspring of the first one and this is how the classification is determined. The livestock of a cohort are classified further according to color configurations, shapes of ears, color of eyes, facial appearance, height, structure of bones, and the sound made by each animal. Turkana herdsmen assign names to individual livestock and will match the stock's sound with its name. By listening to the sound of your livestock, you can recognize your stock from a distance and at night.

Knowledge about livestock herding is learned from the time the child is born and continues into adulthood. The elders explained that from preschool age the child accompanies the adults and the livestock. Children learn by being present, by watching the adults, and by observing every small thing adults do to the stock. The learning is hands-on; children participate in the milking, skinning, and treatment of livestock with the help of adults. They develop a connection with the livestock from the time a stock reproduces a young one. The children master how the kids of the five species of their livestock mature, and study all the physical and behavioral characteristics including reproduction patterns. Specialization begins to emerge in the children as they become older. Skills begin to differentiate in the children and separate into roles such as cattle herders, goat herders, or camel herders.

As the children specialize according to their interest in particular species of livestock, other skills are also learned—knowledge of livestock anatomy, reading the intestines, slaughtering and skinning, milking, extracting blood from an animal, tracking, and breeding. The father and the mother are critical in the training of children. The two parents are responsible for day to day instruction of livestock herding. "Take the herds to this particular water source and graze them in this pasture after they have had the water," or "do not hurt the livestock and be watchful for predators," parents may instruct the young herdsman.

The mother and the girls are responsible for giving the livestock water at the well dug at the river bed. A family can own a water well if they have participated in its exploration and digging at the nearby river beds. Sometimes a well can be a source of conflict if another family uses it.

The parents and older siblings provide guidance to young children. Training continues as parents continue to monitor indicators of competence in children. The support and guidance from adults is gradually withdrawn and children slowly acquire the autonomy to herd. But adults continue to maintain surveillance with a watchful eye indirectly. For example, when the animals return, several indicators are monitored by the parents—are the herds well fed, have they been given water, are all the livestock back, are the children able to answer any technical questions asked? Other questions include: did the herdsman select the right grazing area and was the herdsman watchful on the predators? The frequency with which the stock is lost is an indicator of an incompetent herdsman. Girls and boys are trained in the same way. Girls look after the herds just as well as boys. The mother may take the herds to pasture occasionally but with the highest degree of surveillance. The training of the mother is very rigorous. Even though the work of herding is given to the young people, adults will be available to give support where necessary. For example, when danger is sensed or when an animal is lost adults will be handy to give support to the young herders. The decision to change the area of grazing or to relocate to a different location rests on the parents.

## Water exploration

Water is critical to the herdsmen relying on livestock for survival. When the community is confronted with drought challenges, for example, when traditional water sources dry up, elders have to use indigenous knowledge of their ecosystem to examine and explore alternative water reservoirs underground. The elders study the sand carefully, observing the type of sandy soil on the river courses. White clear sand, grey sand, and black fertile agricultural soil observed on the river bed, or by the dry river banks, indicate the possibility of a water table being close underneath. The elders study the characteristics of certain types of plants, and the places on the river banks where these plants grow have associations with the closeness of a water table to the surface. Generally, places along the river where some acacia woody trees appear green for a sustained period of time will be an indicator of water under the surface. The elders study the rocks underground or at a river bend, they observe the leaves of the trees, they study the frequency with which birds and wild animals hang around at some isolated places along the river courses. Elders study the slope of the land and examine places where land flattens at the bottom of the sloping landscape including river bends. A place where an anteater has dug when it is searching for water is almost the surest way the elders use to tell the presence of water at a dry river course. These remarks were recorded from the elders:

> We test for the presence of water using a spear. We use the tip of the spear to get the indicator that there is water underground. You spear on the sand on the river. The father does this in the presence of his children (girls and boys). When you spear, the spear goes down through soft sand much deeper. When you pull out the spear,

the tip appears wet showing the presence of water. Digging a well is not just the preserve of a particular group. Everybody does the digging.

*(Ng'asike, 2010, p.121)*

The government water engineers use the knowledge of the elders in identifying possible sources of water in the pastoralist areas. Usually where the locals get water, the government water experts are likely to get water from the same place, but in large quantities. Local water experts believe that the knowledge was given to them by God and argue that the knowledge will be passed to their children.

## Fire making

Fire symbolizes the presence of human life in the village or in a home. Besides its use for cooking, fire helps the livestock to trace the home. It is used for branding livestock as well as for treating both humans and livestock. A woman giving birth should stay in a hut lit with fire to symbolize life as well as lighting the environment of the baby. Rituals performed for the infant are carried out with the help of the fire. For example, women and close relatives visiting the mother will chew tobacco and spit some in the baby's fire. Turkana make fire from rubbing two sticks on dried donkey dung. The two sticks are obtained from trees known for producing fire. The locals refer to the two sticks as male and female (Marianna, 2010). Men as well as women can make fire from friction caused by rubbing the two sticks. A family carries two fire sticks whenever they move to new locations. The sticks are safely kept in the house when not in use. Fire is made by spinning one of the sticks with a pointed end into a small groove made on the second stick. The pointed stick is pushed into the second by spinning it hard with both hands into the groove of the other. Human power and energy is required to generate the friction needed to produce fire from the two sticks and for this reason two men or women spin alternately until a small fire ball drops. The fire ball is caught by the dried donkey dug which immediately turns into smoke as the herdsmen continue to blow on it to produce the flames. Fire making is not an activity that is done on a daily basis. It is done only when the home settles in a new place which has no home close by to get fire from. Herdsmen hunting will need fire to roast their prey, and fire has to be made from dry tree sticks. Otherwise, traditionally fire continues to exist in the home as the family has the responsibility for ensuring it is not extinguished after it has been used for cooking.

## Story-telling and folktales

In addition to the narratives of the culture that children receive as they participate in or observe family household routines, pastoralists may spend time in the evening educating the children about the cultural beliefs, norms, manners, moral values, and family traditions using folktales. The stories are very entertaining and very insightful. Children tell these stories to teachers at school during the mother tongue lessons. In some of the schools, teachers have recorded these stories in print, published them in short story books, and kept the collections in the library for children to continue reading from early childhood and lower primary grades (Ng'asike, 2017). A few of the stories the author collected and shared with children who were visited in schools are listed below in Table 5.1.

**TABLE 5.1** Short stories collected from children

| Turkana title | English title |
| --- | --- |
| Ekilam Aito | The Danger of Abandoning a Mother |
| Ebele Ebu Along'or | Greedy Hyena |
| Ikoku nia Ayauni Ekisil | The Child as Peacemaker |
| Ekerikan | The Hunter |
| Nyo Kolong Ayauni Anam Turkan | The Origin of Lake Turkana |
| Ajokis Eng'ol | Benefits of Palm Tree |
| Namur Tung'a | Stone People |

## Conclusions and implications for curriculum implementation in pastoralist communities

This paper argues that children have knowledge acquired from their cultural capital that needs to be recognized and integrated into formal schooling (Krätli, 2001). The curriculum in formal education basically has content that is very similar to the content of the indigenous traditional education of the everyday life experiences of children. For example, the calendar, livestock husbandry, plants, universe, just to name a few examples appear to be content of the curriculum that both formal and traditional education value. However, Turkana children will finish school without hearing their own meaning of a calendar being acknowledged in the formal school settings. Turkana children are vastly knowledgeable about livestock herding skills, which include tracking livestock, treatment, and reproduction, including the psychology of animals. But formal curriculum, in the case of Kenya, operates in total disregard of the cultural capital of the children. This is the challenge the majority of African children face in modern education which operates as though education is an importation from cultures outside Turkana.

## References

Eromose, E. E., & Danny, M.S. (2017). "We know our terrain'": Indigenous knowledge preferred to scientific systems of weather forecasting in the Delta State of Nigeria. *Climate and Development*, *11*(2), 112–123.

Krätli, S. (2001). *Educating Nomadic Herders out of poverty? Culture, education, and pastoral livelihood in Turkana and Karamoja*. Brighton, UK: Institute of Development Studies, University of Sussex.

Lave, J., & Wenger, E. (1991). *Situated learning: Legitimate peripheral participation*. Cambridge, UK: University of Cambridge Press.

Maanen, J. V. (1988). *Tales of the field: On writing ethnography*. Chicago, IL, London: The University of Chicago Press.

McCabe, J. M. (2004). *Cattle bring us to our enemies: Turkana ecology, politics, and raiding in a disequilibrium system*. Michigan, MI: University of Michigan Press.

Malinowski, B. (1922). Introduction. In *Argonauts of the Western Pacific: An account of native enterprise and adventure in the archipelagoes of Melanesian New Guinea*, pp. 1–25. London: Routledge & Kegan Paul.

Marianna, B. (2010). *The children of Eve: Change and socialization among sedentarized Turkana children and youth*. Master's thesis. Norway: University of Bergen. Retrieved from https://bora.uib.no/handle/1956/7382.

Ng'asike, J. T. (2010). Turkana children's sociocultural practices of pastoralist lifestyles and science curriculum and instruction in Kenyan early childhood education. PhD Dissertation. Arizona State University.

Ng'asike, J. T. (2017). Implementing early childhood development and education curriculum in nomadic pastoralist communities in Kenya. *AnLeanbhÓg—The OMEP Ireland Journal of Early Childhood Studies, 11*, 127–135.

Ng'asike, J. T., & Swadener, B. (2015). Turkana indigenous knowledge on environmental sustainability and pastoralist lifestyle for economic survival. In E. Sumida Huaman & B. Sriraman (Eds.), *Indigenous universalities and peculiarities of innovation*, pp. 107–127. Rotterdam: Sense Publishers.

# 6

# THE CALL TO NATURE PERMACULTURE PROJECT

*Lolagul Raimbekova and Solomon Amuzu*

Permaculture is "a philosophy of working *with*, rather than against nature; of protracted & thoughtful observation, rather than protracted & thoughtless labor; of looking at plants & animals in all their functions, rather than treating any area as a single-product system" (Mollison & Holmgren, 1978). Although the original focus of permaculture was sustainable food production, the philosophy of permaculture has expanded over time to encompass the ways in which larger economic, social, and political systems are intertwined within local models of production and preservation. The story of Solomon Amuzu's Call to Nature Project (CTNP) mirrors these intricacies. Although promoting sustainable, school-based gardening is one of the overarching goals of Solomon's CTNP project, the socio-political networks that have evolved in Amuzu's work are indicative of the kind of complex, local, and less-local interdependences that operate within permaculture movements. The following story was constructed from semi-structured interview data generated between the first and second author in order to highlight the ways in which one educator is working to sustain children, elders, and earth. Understanding the experiences of this activist educator—from the beliefs about nature that were shaped by his early childhood to the pragmatic steps taken to influence educational practices at a local level—may help others who wish to leverage community resources to bring sustainable change to their own educational contexts.

## Hunger, humility, and connection

I was born and raised in the small village Brekumanso in the eastern part of Ghana, a country in West Africa. My family is big and consists of nine people, where I am the second child. The community in which I grew up was mainly involved in agriculture and farming, and most of my close and extended family members were farmers and peasants. I was involved in taking care of the land and agricultural work from my early childhood. The land that my family and I worked on was our second home, as it not only gave us food and helped us to survive the difficult times of hunger and drought, but it also gave us enjoyment and peace of mind. Also, always being outside as a child I developed a strong sense of connectedness to nature. In other words, nature for me is not something out there, it is a part of myself.

Once, due to hard life circumstances we moved to the capital city Accra, but living in a high rise building I felt so disconnected from nature and all its blessings that I decided to come back to the village and live in harmony with nature and all its glory again. Though working and cultivating the land requires a lot of hard work, it gave me a lot of enjoyment and I was always curious about what other kinds of crops or vegetables could be planted in the lands we were working on. Some of the main crops we grew were cassava, plantain, cocoyam, banana, cocoa, and yam. Many people reading this might think since we were growing a variety of crops we were making enough to provide for the family, but the answer is no. We were operating on a system call *Do Tse* (meaning share of harvest); since my family owned a small piece of land it was not always enough to feed a big family. Therefore, we needed to work on someone else's property and at the end of the season the produce was divided into three parts—one part went to the landowner and only two parts left for us. Sometimes getting access to three square meals a day became a challenge due to the high level of poverty; sometimes we had to eat without any protein source (this condition remains still among some villages as of 2017).

I do recall times that we enjoyed a good meal only when we were lucky enough to gather snails during the rainy season, harvest wild mushrooms, or if the father of the family was able to trap bush animals such as rat, grass-cutter, squirrel, or tortoise. Sometimes I had to join a class mate to run to the bush and chase bush animals with dogs to provide a meal for the family. Only once I reached the age of 24 did I realize all animals have the same right to live as we humans do, but somehow this right is not even considered when it comes to food.

## My call to nature

I guess I was lucky—despite the hardship of life and poor education in my high school years, I was able to attend the university and get my higher education degree, majoring in agriculture. My luck was not only due to my hard work but the generous support and continuous inspiration of an American Peace Corp volunteer named Kate Schachter, whom I always describe as a friend, a mother, and a mentor. Additionally, there was an Australian teacher in Ghana, Gregg Knibbs, who introduced me to the idea of permaculture. Learning about permaculture, the development of an agricultural ecosystem intended to be sustainable and self-sufficient, it seemed as if I developed "new eyes" for seeing the world. This defining moment was when I found my *"call to nature."* I think I was "called" by Nature to care for Her.

## Permaculture's promises

My first concern appeared when I started to realize and witness with my own eyes how modern agriculture is causing harm to the land and people through the use of poisonous chemicals. Therefore, I was motivated to look for alternatives, and to bring the organic features of the land and the produce to the forefront. I was always thinking about developing a project that was not only able to bring life to the land, producing organic food, but could also serve as a teaching tool for teachers and as a learning tool for children—helping them lead happier and healthier lives in a rapidly changing world, making a positive contribution to society.

My second concern came from my own schooling experience; as students we were given a critical mass of information, never gaining any idea about how to use that information and how to apply it in practice. We had very little communication in school, and were not engaged as much with the subject. Additionally, we never had to express our opinions or ask questions, and of course, the curriculum rarely mentioned nature. With that in my mind, I strove to move forward despite all the hardship I faced, and eventually launched the not-for-profit organization Call to Nature Permaculture (CTNP) project in February 2015 to provide an alternative. My immediate action was to help my community, driven by the wide array of everyday struggles and environmental issues we are facing in Ghana. With my two committed colleagues, Samuel Badu Adotey, our Deputy Director, and Rachael Yussif Ramatu, our secretary, I started the project with three goals in mind: 1) A grey water system to improve water management for greater crop yields; 2) school gardens that would engage schoolchildren and ensure they had access to healthier, organic produce; and 3) more trees, which have been planted to fight climate change and soil loss, and increase wildlife habitat. The mission that we have for our project is to care for the Earth, care for people, and share resources by implementing permaculture principles and providing a high quality of permaculture education and training. We envision a world where every student helps to maintain and take care of a garden in their respective schools, every citizen helps to maintain and take care of a garden in their community, and because of that everyone has open access to organic and nourishing food.

## Difficult first steps

Learning more about permaculture, I started to understand that there is an interconnection between man, other living things, and non-living things. Each element in a system can support every other one, based on the way we place them. There is no such thing as waste when you talk of permaculture. For example, the chicken system, fish pond, and plant garden have a relationship to play as a whole design structure. The chicken coop can be placed on the pond where the chicken droppings feed the fish, and the fertilized water in the pond feeds the garden. This is a complete system that saves money and time.

We started our permaculture experiments on four acres of land, donated by a local farmer on lease for a period of four years. Certainly, only leasing land is not sufficient to promote permaculture products over the long term; thus we are in constant need of financial support. While the fundraising process to support the project was going on, I nursed more than 10,000 trees on free land in one of the local schools and involved volunteers from the community to plant them around the schools in Accra, the capital city, and Aburi, a city in South Ghana. This project served as a starting point and was undertaken without any financial support, but mobilizing community resources helped to a big change to come about. In addition, this project served as an introduction to our permaculture project in the local region, since we wanted to come back to schools with a more important mission over the course of time.

## Building alliances

Bringing this model to the local schools I wanted both teachers to have an alternative for their in-class teacher-centered approach and children to be exposed to a more hands-on, minds-on learning, spending time outdoors and reconnecting with nature through education. Creating opportunities to reconnect with nature is crucial for both children and adults, if we want to protect our environment and biodiversity. We all need to spend more time unplugged from our electronic gadgets and find better ways to let nature balance our lives. And this is what I am striving for in my project.

Unfortunately, the education system in Ghana does not promote the required education or healthy lifestyle and lacks the economic and governmental infrastructure to do so. The poverty level is so high that you need to carry your own chair and table in and out of class every day—this is in addition to the food insufficiency and malnutrition. There are many instances when students sit on the bare ground or under a tree to study without having proper meal for a day. Therefore, one of my dreams was always about how to support schools in developing programs that could support not only the provision of healthy food but also the education and skills for how to grow the healthy food at the same time. Moreover, I hope that being involved in gardening children become stronger, more active learners that can think independently and use their knowledge and skills to solve the challenges they face in school and in their future life. I also hope that children embrace a healthier lifestyle as an important tool to achieve success and become more resilient, confident, and responsible students.

## School nutrition and outreach

In 2013, before launching the Call to Nature Permaculture project we conducted a small study to learn about the government food supply to our local schools. Our findings revealed that the Ghana government school food supply mainly contains carbohydrates, and lacks sufficient nutrition. Based on these findings we decided to improve the situation and try our best with the idea of sustainable food production using permaculture to provide and enrich the school food supply with more nutrition. Once CTNP project was under way in summer 2015, the first project school we started to work with was the local school, known as Oyarifa Basic School, in Accra, Ghana. This school offers preschool, elementary, middle, and junior high school education, consequently covering students ages 3 to 4 years old (preschool), up to 16 to 17 years old (high school). In our project, we targeted the school as a whole community involving students from preschool up to high school along with teachers, administrators, and other support personnel.

We, my project team and I, wanted to also introduce the idea of working together as a community—not only working with the specific grade. I know the idea of school as a community is very well known in the Western world, but for us here in Africa it is a new concept, in particular bringing people of various ages together. The school, Oyarifa Basic, was selected based on its proximity to our main project site, and based on the idea that the school has its own piece of land to plant a sustainable vegetable garden, which students can monitor and cultivate over time on their own. The connection between CTNP and the independent space at Oyarifa provided a springboard to give us a possible model for other schools. We were encouraged.

Through the project, the students and teachers learned how to grow various plants using permaculture methods in our main sites first, then later they transferred the knowledge from our site into their own garden in their school. Daily, more than 500 children aged 3 to 17 spent five hours of their school time in the gardens, learning and practicing the implementation of organic farming. The time they spent in their garden were the hours of their science classes. Although the curriculum was not officially modified to meet these changes, the school administration was highly supportive of the idea and allowed the project to grow and expand.

## Mentoring teachers

Since teachers find this project innovative they are part of the learning process as well. My team spends more time with teachers (around 8 hours per week) to help them understand permaculture, build concepts, and find ways this learning can be continued in the "official" class time. For instance, if during the week in the field both teachers and students learned about growing broccoli, kale, and garlic, during in their "in-class" time they would continue their exploration about the topic to better understand the process.

One of our essential goals of working with the school community is to enhance the understanding of students and teachers about how eating healthily benefits the body and how growing healthy food benefits the earth. I think a quote from one of the students supports this: "I have never eaten greens before and never saw how to grow them. Now I have learned about the benefits I will try to eat them more."

One of the main activities we carried out with the schools was teaching students how to grow and take care of their own garden, as well as teaching and learning about the benefits of gardening and the long-term outcomes, both for the school and the country. Moreover, we helped and taught the students how to plant Moringa trees all around their school.

Children, from the very young and the older ones, were involved in every step of the process. While working on this project, they learned about Moringa. Moringa is a single tree that has three times the potassium you would find in a banana, four times the vitamin A found in a carrot, and seven times the vitamin C found in an orange. Moringa is also rich in minerals, vitamins, essential amino acids, phytochemicals, vegetable proteins, anti-oxidants, anti-inflammatory agents, and carbohydrates. As a result of the hard work at the Oyarifa Basic school, our main project school, we, as elders, were able to use it as a model school for other new schools involved in the project.

## Managing resources, technologies, and global partners

Since we started the project on a very limited budget, our first task was to raise money nationally and internationally, applying for various grants, using various social media avenues. We tried to spread the word globally and received financial support from many organizations, both nationally and internationally.

Our first source of funding begun with the Go Fund Me page on the internet that enabled us to raise around 2,000 USD out of 6,000 USD. With the help of that funding we procured farm tools, seeds, watering cans, and water storage containers, enabling us to start our project activities. In addition, some other organizations that supported our mission are the following:

*Returned Peace Corps Volunteers (RPCV)*, a group of returned Peace Corp volunteers from the United States that support our school garden project through their Gift Away grant to the amount of $1,000 each year.

*The Pollination Project Foundation*, a seeding project that change the world, is a nonprofit organization based in the US that gives seeds grants to social change leaders. We have been in partnerships for the last two years and received their grants to support our projects.

*First United Methodist Church*, based in Colorado, US. This mission church has provided a small amount of money to pay for our farm operations and support the missionary schools in Ghana in the last two years.

*Water charity*, a US based organization, has been our main partner for our tree planting project. Our first project with them was to install an irrigation system on our property to enable us to water our fields more easily and effectively. The second project, enabled us to plant 20,000 trees on the streets of Accra. This year, 2017, we are again planting trees to protect the water body from drought. Our main partnership with this organization is to plant 100,000 trees by the year 2020.

*Food for all Africa*, a local social enterprise based in Ghana that supports our mission and we work together to supply the schools, orphanages, people with special needs, homeless people and elderly in our community with the nutritious food.

In addition to that support, CTNP has received many foreign visitors and students from the University of Wisconsin, Madison, USA, Germany, the UK, Nigeria, Kenya, Togo, and Switzerland who are interested in learning from our experiences and disseminating the knowledge into their respective countries and societies. I have to mention that I am highly grateful and blessed to have developed strong partnerships with both national and international entities that have not only helped my project to grow in Africa, but have also shared the word worldwide.

## Educational and worldly aspirations

Our work at CTNP is just getting started and our dreams are coming true. But for our team, achieving one dream does not mean not to be inspired by "new" dreams. In this short period of our project implementation, not only have my devoted team and I created gardens for schools, but schools have also begun to share our skills through permaculture education. Permaculture is especially helpful to local farmers who are in need of more productive and ecologically sound designs and techniques to add into their farming methods. Recently, for example, my project brought together elders from other parts of Ghana to teach them permaculture principles. We are actively restoring the marginal lands around our community, particularly those degraded through road building projects and other urban impacts. We are planting trees, teaching and modeling methods beneficial to humans and nature through a technique called alley farming along with tree planting. CTNP's team is highly devoted to saving our little brothers and sisters in Ghana, as well as obligated to try our best to create a better and safer world for this next generation. For the past year of operation, our project has been able to feed many schools and other institutions around Accra and the eastern part of the country.

In the past year, our project has given one square meal per day to 600 school kids (age 3 to 12), 150 orphaned children (age range from 4 to 16), and 50 disabled elderly people (age 45 to 65). This is how we distribute the harvest from our project vegetables: 50% goes to the project beneficiaries, we sell 30% to pay for our operation expenses, and we donate 20% to orphans and people living with disability. Some of our beneficiaries are Kunkunuru Methodist Basic School, Oyarifa Basic School, Osbin International School, Gyankamah Methodist Basic School, Sun Shine Academy, De-Best International School, Ayim L/A school, Frafraha Childrens Home, Jaynii Orphanage, and Akropong School of the Blind. With expansion, we plan to feed about 3,000 people in 2017, 5,000 by 2018, and 10,000 by 2020.

We definitely have huge hope for a better future, and some of our dreams are to educate the youth about taking responsibility to sustain the environment, plant their own gardens, learn more about the ethics of permaculture, and simply become a generation with healthier eating habits as well as being responsible citizens who care for our Mother Earth.

## *Providing children with a meaningful relationship with nature*

Gaining a better understanding about the permaculture concept, I became convinced that earth care, people care, and fair share (the foundational ethical principles of permaculture) can be part of the solution to all the environmental challenges we face today in the world. In addition, I cannot emphasize enough the benefits of school gardening for teachers, students, and communities. To name a few:

- Active learning and student engagement
- Student attention and class management
- Developing/training teachers as gardeners
- Connection to nature, history, and community
- School pride (like a team of sport)
- Academic achievement
- Social, emotional and physical health (and many more).

But the most important thing is to pass this knowledge to the children and develop this project in a way that it can serve not only as an alternative to the in-class education in Ghana, but also as a model, become part of the national curriculum, and extend in-class education to more practical and hands-on learning where children feel connected to nature and lead their own learning. In fact, from my own evidence of working with school personnel and children I suggest gardening plays an important role in the development of a child. As an educator, I believe every child should have the chance to experience the benefits of this learning. I am committed to spreading the word further and to championing the importance of gardening in every school in Ghana. My message to educators around the world would be to inspire and support your schools to embed gardening activities in the heart of your school life and maybe we all still need to find the best ways to educate our youth.

## Reference

Mollison, B., & Holmgren, D. (1978). *Permaculture one: A perennial agriculture for human settlements.* Tasmania, Australia: Tagari Publications.

# 7

# THIS IS MY DAD AND HE'S A SCRAPPER

## Curriculum, economics and clout in kindergarten

*Jonathan Shaw and Katy Morgan*

## Introduction

This chapter reflects on the experience of the first author as he attempted to disrupt traditional power dynamics by engaging in negotiated curriculum (Hyun, 2006) in a kindergarten classroom by welcoming the knowledge base of a parent with little institutionalized cultural capital or clout (Bourdieu, 1986; Rishel, 2008). This collective ownership of the curriculum centered on the shared experience of learning about the properties and alternative uses for "trash" or waste materials from a parent, while also raising an increased awareness of the environmental effects of human consumption. The work simultaneously facilitated the development of arts-based community expression of the kindergarten students.

At the beginning of the school year I did not know what role trash would have in our classroom curriculum. I also did not know how, because of one student and his father, we would look at trash with a different lens. This chapter examines a project that took place in my classroom involving Keith (a father), Kevin (a son), and myself (a kindergarten teacher). Under an umbrella of Paulo Freire's views towards *historical fatalism*, I use a lens of negotiated curriculum (Hyun, 2006; Kuby, 2013) and clout (Rishel, 2008)—be it financial or social-cultural—to examine how, by disrupting power relationships commonly found in public urban school districts in the United States, which serve predominantly lower socio-economic students and families, teachers can elevate the role of community elders into important decision makers in classroom curriculum.

Clout, as theorized by Rishel (2008), is essentially power and the ability to influence the powerholders of a system. When applied to education and the school system, clout can appear in several forms. For the experience discussed in this chapter, two types of clout are particularly significant: financial clout and social-cultural clout. Education is, in many respects, an art with mimesis of the larger society occurring within the school, forming part of the hidden curriculum that helps shape the experiences of every student. Financial clout in the school reflects the disparity of socioeconomic power held outside the school—parents and students with little financial holdings also hold little power within the school and the intra-school culture. Similarly, social-cultural clout reflects the type of power and influence that people who fit the ideology of the dominant society are able to exert in the systems of the society, including the school. For parents that do not fit the white middle-class conceptions of a desirable education, employment, or general community status, a "scrapper" for instance, they are very unlikely to be perceived as valued contributors and decision makers in the school.

The issue of power highlighted inequities found across the various stakeholders of educational systems resonates loudly with the work happening in this chapter. What Hyun (2006) and Rishel (2008) illuminate is the idea that the marginalized parts of a community have little to no power over what curriculum looks like in the classroom unless teachers take a critical lens towards practice (Bourdieu & Passeron, 1979; Freire, Araujo Freire, & De Oliveria, 2014). When allowing for a disruption of power by repositioning clout in curriculum decision making, an organically emerging, negotiated curriculum can develop within a classroom setting. At Patton Elementary School, many of our stakeholders—the students and their families—are members of marginalized portions of our society. By paying closer attention to a movement created by local artists and city government, along with my student's lived lives, I let go of complete control of the curriculum to perform what Hyun calls "negotiated" curriculum, to impart ownership of the curriculum and our shared educational experiences to our community. This led to a year-long curriculum project revolving around scrap or "trash," art, and reuse of objects commonly thrown away. In this instance, trash re-emerges as repurposed and people commonly marginalized are repositioned socially, disrupting larger systems of politics and governance controlling public education towards a more critical eco-conscious curriculum.

## Why negotiated curriculum: power holders, historic fatalism, and the market

Socioeconomic status comes with inherent power or lack of power when teachers make decisions in the form of curriculum pursuits in the classroom (Bourdieu & Passeron, 1979; Rishel, 2008). The powerlessness of lower socioeconomic classed parents is well documented (Doucet, 2008; Lareau & Horvat, 1999), and a lack of cultural capital further removes parents and their knowledge from contributing to the dominant curriculum or being validated by the enacted dominant curriculum. Further, Freire, Araujo Freire and De Oliveria (2014) illustrate the idea that fatalism, a belief that we are powerless to fight against the predetermined course of events, operates in educational circles in regards to the economic market. Essentially, in the context of the United States of America, capitalistic values push society towards "a dehumanization of education" (Freire et al., 2014, p. 69) in which the sole purpose of education is to implicitly produce more consumers who place value on their lived lives through the ability to climb within socioeconomic statuses and material accumulation. Without a rejection of the power structure, the system remains the system, students become consumers, and we all perpetuate the dominant ideology through a system of education dependent upon the capitalist inequities in society.

It is desirable, in a democratic system, to construct the curriculum in a manner that reflects the ideas, opinions, and goals of the participants. Hyun (2006) offers an idea of "negotiated curriculum" as an understanding that there are powerholders of curriculum in the classroom, which includes recognizing the important role of the students and their parents or guardians. Enacted, negotiated curriculum includes the children and is also, importantly, directed by the children and the context (Kuby, 2013). By being aware of the power dynamics in the classroom, teachers can work towards a more democratic curriculum by sharing curricula decision making in the classroom with students and families.

During my first year teaching at Patton Elementary School, I spent most of my time stumbling through packaged curricula that held only the slightest interest of my students and forced them to spend most of their days disconnected from the lives they were living outside of school. Hyun's (2006) work resonated with my plight; she states, "teachers are set out to teach according to a planned curriculum via instruction-orientated teaching, without engaging the interests of the students" (p. 117). As a first-year teacher, I found very little freedom in my practice to adjust the curricula to truly meet the interests and lived lives of my students. However, feeling more confident in my second year of teaching, the experiences upon which this chapter are based, my students and I embarked on a journey that initiated a dance between the external pressures of the district, state, and federal mandates and an inner search to provide for the self-needs of teacher, student, and family in our educational experience. It is vital for this chapter to take time to more closely examine how Keith, Kevin, and my own clout (Rishel, 2008) eventually fit into what might be seen commonly as a fatalistic construct of educational policy decision making in the American public classroom.

## Context and identity

To honor the positionality of each of the co-creators of knowledge in this project, we attempt to provide information regarding the context and aspects of the identities of Kevin as a student, Keith as a father, and the first author as a teacher. Pseudonyms are used for all people and places except for the author/teacher, Mr. Shaw. The narrative of teaching is written from the teacher's perspective.

### The rust belt and Patton school

I live and teach in Northeast Ohio, a section of the Midwestern United States, in an area of country commonly referred to as the "rust belt." This term generated from the color of the factories and mills left vacant after the majority closed and were left to rust as a visual reminder of the once prosperous industry. From the late 1800s through the mid-20th century was flush with steel mills, automobile and tire factories, and coal mines. Once vibrant and powerful engines of our local economy, the abandoned mills and factories remained as hollowed memories across the landscape. Patton is located on the southeast side of the city close to a scarcely running steel mill. Many of the houses in the neighborhood are either run down or abandoned. Driving down streets close to the school, one can see the evidence of economic hardships endured by the community: houses with boarded up windows, porches falling off, or collapsing roofs.

Patton, like many other public schools in the US, had adopted what was known at the time of this writing as standards in "The Common Core" of Mathematics and Language Arts (National Governors Association Center for Best Practices, Council of Chief State School Officers, 2010). A standard in this context referred to a topic or concept that must be taught to and mastered by students at a specific grade, spanning preschool (3–5 years of age) up through high school (grade 12). These standards were created with the intent to standardize education in mathematics and language arts on a national level. Our own state, Ohio, created sets of Social Studies and Science standards to be covered in the k-12 public school classrooms. Federal and state funding for public education have been tied to the implementation and evidence of students' mastery of these federal and state standards, documented with a series of standardized tests and assessments. In Ohio, teachers have to use what is known as researched based, data driven curriculum in the classroom for instruction. In many school districts, administrators rely on curriculum companies to develop curricula that fits federal and state requirements of standards in the classroom.

One attribute both the state and federal government use for a statistical measure of schools is the economic status of families. How many families living under the national poverty level determined by the federal government decides the level of state and federal financial assistance each school district and school receives. The percentage of families at Patton school who lived or were labeled as under the national poverty level varied from 80% to upwards of 90%. Families of Patton Elementary vary in their structures and makeup; most families consisted of a single parent with two or more children. Adults in families were either married, divorced, or never married, and many families had step-parents—both formal and informal. Other families consisted of grandparents raising their grandchildren for a variety of reasons, such as parents not being able to financially support their children, incarcerated parents, multi-generational households, and so on. For the children of Patton, the collective elders in the families played different roles within the structure of each family while serving as caregivers to the individual students.

### Learning about Kevin as a student, Keith as a father, and reflecting about myself as a teacher

Kevin, one of the kindergarten children in our classroom, was a loving child who spoke excitedly about caring for his family, especially his baby sister. He split his time between two houses; his mother and father were never married and had broken up before Kevin enrolled in school. He maintained a loving relationship with both parents. When Kevin was with his mother, Teresa, he shared an apartment with a half sibling and two step-siblings from his mother's partner. Teresa's main source of income consisted of government-funded public assistance—commonly referred to as welfare—and she purchased food for the household with food stamps (another form of government-funded public assistance). Teresa had not been working for over a year, and her most recent employment was at a local fast food restaurant. Kevin's mother, and Kevin by extension, held little financial or social-cultural clout as judged by a white, middle-class dominated system.

Keith, Kevin's father, identified himself as unemployed, but exhibited resourcefulness in his efforts to support his family outside of formal employment. He lived in an apartment with his girlfriend and two of Kevin's half siblings, and was expecting a child with his girlfriend to be born during Kevin's kindergarten year. While he did not discuss much about his personal life, Keith spoke about his frustrations with the school system, and his perceived lack of power within the system, both as a former student and as a current parent to students. Keith reported feeling that no one (i.e. powerholders within the school) listened to him when he was a student, and that no one was listening to him as a parent. He had other children struggling academically and found parent–teacher conferences frustrating because they only talked about his children's test scores and behavior problems. Keith expressed frustration because he felt stuck in a system that was working against him (via his teachers as a child, and his own children's teachers). He wanted to be able to support his children's teachers and stressed the importance of getting an education, but was very discouraged because he felt as though his children were up against all of the same struggles he faced as a child—the lack of possessing or even gaining social-cultural clout

As a teacher, I also reflected that my parents divorced when I was very young and each went on to be remarried. I lived with my father, stepmother, and two brothers. My family's socioeconomic status was working class by American standards, my father holding a handful of different jobs and positions and my stepmother working as a hairstylist and owning her own salon. In all, I was provided a comfortable life, never exposed to difficult realities such as not having enough food in the house, utilities being shut off due to not being able to afford the bill, or eviction of our household because we were not able to pay our mortgage. While in high school I realized that if I wanted to go on to college after I graduated, I would need to provide my own financial means to pay for tuition. I worked in factories, drove a semi-truck, worked in construction, made custom furniture, and taught woodworking in a local wood shop store. These efforts were all in the hope of being the first person from my immediate family to obtain a Bachelor's degree from the state university. In my family, having a college degree was one of the only means to get "ahead," both financially and socioeconomically.

All three of us—Keith, Kevin, and myself—fit in the larger societal constructs of American culture, influenced by pervasive ideologies of power, education, and capitalism. In the remainder of this chapter I will share the story of Keith and Kevin through the work of a year-long project taking place in our classroom, deemed by the students as the "Art Walk Project." The main provocation for this work started on a crisp and sunny Saturday in October where I took my students and their families on a field trip to the art district found in Patton's city.

## "Art Walk Project" and the "junk" art

The city in which Patton our school is located was once an economically booming city benefiting from auto and other manufacturing industries in the area. Now the large buildings in which these industries were housed sit idle from mass shuttering of factories. Leftover scraps of the past industries remained within the abandoned factory buildings. In a move to "reinvigorate" the downtown area, the city government decided to begin funding art installations around the city center. The employment of local artists and craftspeople lent public interest to an already strong food and restaurant trade in this area. Many murals, scrap metal sculptures, and other art media began to spring up all around downtown. Every day during my drive through downtown to Patton, I would pass the most amazing sculptures made from the remnants left over in the abandoned factories. In the middle of downtown, in what is known as the "Arts District" sat a large giraffe, sea turtle, bear, fish, and saxophone all of which were made from scrap metals. One sculpture, a rhinoceros, had been made from scrap tires (see Figure 7.1).

It was the middle of September, early in the school year, as I walked around downtown taking pictures of the large sculptures. My intent was to bring these pictures to a classroom meeting to share with my students because I found them so fascinating. Unlike my colleagues in the school who used classroom meetings to disseminate weekly and daily goals related to district mandated curriculum, I used our classroom meetings to talk about topics of interest that the students or I wanted to share. The first day I brought the pictures to our meeting, the class erupted with excitement and stories. Unknown to me, the sculptures were well known by my students and their families. Through our classroom meeting I found out that on the first Friday of every month, the city hosts an art festival where streets are closed down and city residents are invited to enjoy the art created and being created downtown. It is during these "First Fridays" that new art installations would be dedicated and installed in the "Art District" downtown. It was at this meeting the students decided that they wanted to do something with these sculptures.

**FIGURE 7.1**   Giant giraffe and monkey sculpture and tire rhinoceros
*Source*: Jonathan Shaw

The students and I contacted a local artist, Mat, who worked with the sculptor of the scrap animals downtown. Mat agreed to take our class on a guided tour of the "Art District" to tell us about all of the artists who created pieces of art out of scrap materials. On a sunny, crisp Saturday autumn morning most of my students and their families met "Mat the Artist" for a guided tour. During our tour students and families interacted with the scrap metal sculptures postulating ideas on how and why the artist made such creations. After our tour, the class met in the large common green space for lunch. Inspired by their time out of the classroom, experiencing the beauty of the arts district, the students decided that they too wanted to make a sculpture using trash and scrap materials. After our lunch on the green, some students and families went home and other students and families accompanied Mat to his art studio for a tour. While visiting the art studio, students learned how Mat and fellow artists collected, sorted, and stored the recycled materials used in the creation of their art.

Once back at school the following Monday, the students created a plan to work with Mat building a sculpture. A polar bear, being Patton's school mascot, was chosen to be the final form of their sculpture. We wrote letters to our families explaining our need of recyclables and included itemized lists of materials needed: milk jugs, aluminum beverage cans, metal food cans, and anything else useable for constructing. Letters to families were sent home and we began collecting materials for *our* polar bear sculpture. The classroom became a staging ground for found items, recycled materials. The students, remembering their visit to Mat's art studio with his art materials sorted in large bins, began self-guided sorting activities with their newly found art materials, exploring the physical attributes of the objects. Keith was about to extend our sorting activities and help the students become more aware of issues around trash and recycling.

## Negotiating the curriculum and Keith's trash

Our school district typically hosts two parent-teacher conferences a year. The purpose, set by district administration, was to review student test scores with parents and address any other classroom concerns teachers had. The first of these parent–teacher conferences is held in November, the second in February. Keith arrived at his scheduled parent–teacher conference with a broken lamp, a magnet, and an old brass apple. I was very intrigued as he walked into our classroom and sat down at the table beside me and laid his objects out next to Kevin's district test scores and progress report. Eager to learn why he had brought this collection of items to our classroom, I pushed the reports, tests, and traditional power dynamic to the side and listened.

> Mr. Shaw, my son came home telling me that you guys are trying to collect recycled materials to work with in the classroom. I don't know if you knew this but I am a scrapper. I go around each week before the trash gets picked up and collect all of the metal I can find. I take the metal to the scrap yard and turn it in for cash. I haven't had a full-time job for a while and this helps us get by with food and all.
>
> *(Kevin's Dad)*

In this instance, it became clear to me, that in a conventional sense, Keith, Kevin's father had two significant types of clout working against him, financial clout and social-cultural clout. Keith, having not completed high school and lacking a desirable job or home, lacked the social-cultural clout held by many parents and all teachers at Patton. Every teacher in the school at minimum held a bachelor's degree, including myself. Being an unemployed person, but benefitting from governmental assistance through his girlfriend, Keith would have also been perceived by most as lacking financial clout among the middle-class population of Patton's staff. Looking past these limited views of power and renegotiating curriculum with the community transformed how Keith could be viewed socially. Kroeger and Lash extend this idea of reversing power dynamics between institutions and families even further, "[w]hen the parent is viewed by the teacher as one with knowledge and power with experiences and perspectives to offer rather than an individual to be coached or changed, the power dynamic within the institutional relationship is shifted" (2011, p. 270). At this point, Keith put the broken lamp on one end of the table and the brass apple bell on the other. He had taken the extendable magnet in one hand and began to demonstrate:

> Now I have to get a lot of metal to really get any kind of money, I know what you are thinking, I use this magnet to pick up pop cans, but you'd be wrong. Did you know that there are two different kinds of metal the scrap yards are looking for? There is ferrous metal that is magnetic, like iron and steel, and non-ferrous metals, like aluminum and brass, which is not magnetic. The scrap yards pay more money for the non-ferrous metals. See, when I take the magnet and hold it to the apple, it will not stick. That tells me that this metal is non-ferrous and I know that I will get more money for it. Now when I take the magnet and hold it up to the broken lamp pipe it sticks, so it's not worth as much money. Kevin came home excited because you asked everyone to start looking for trash and recycled materials and he knows that's kind of what I do. I go around and pull metal out of people's trash and I recycle it at the scrap yard. Kevin knows about how I use the magnet to find the more expensive metals and wanted me to come in and teach you about it. I wanted to give this stuff to you guys so you could use it in the classroom.
>
> *(Kevin's Dad)*

In a typical traditional classroom, Keith's overall lack of economic and social-cultural clout would have prevented this interaction from ever happening. However, by rejecting common perceptions of clout, this parent was repositioned in the curriculum as a power holder, with learning and teaching becoming more balanced toward things that mattered to the parent the child and the teacher (Hyun, 2006; Kroeger & Lash, 2011; Rishel, 2008). During the parent–teacher conference that night the focus shifted. Instead of focusing on test scores and academic progress, my role as the teacher shifted to learner as power dynamics were renegotiated. Keith taught me about ferrous and non-ferrous metals and expressed a great interest in teaching the students of the classroom everything he knew about collecting metal out of the trash. Instead of sticking to district prescribed parent–teacher conference agendas, we planned a learning experience for the students.

## "My dad's a scrapper" morning meeting and project

A week after our parent–teacher conference, Keith was able to come into the classroom. Kevin was extremely excited and proud that his dad was coming in to teach the class. We had marked on our classroom calendar the day Keith was coming and eagerly counted down the days until his arrival. We gathered in a circle at our classroom meeting spot and welcomed Keith. Kevin explained to the class why his dad was there that day:

> Hi guys, this is my dad and he's a scrapper. A scrapper goes around and takes metal out of the trash people throw out. It's a really cool job and I even help sometimes. We will get into my dad's truck and drive around real early in the morning to pick up any metal we can find. Sometimes we even go to the stores downtown and my dad gets into the big dumpsters to look for metal. After we fill up his truck we take

it to the scrap yard, but before we can dump our metal we have to sort it first. We use this magnet to see what metal sticks to it and what metal does not. We want more metal that does not stick because we get lots more money.

Kevin and his father's understanding of the environment and repurposed materials surpassed mandated social studies curriculum foci on consuming and producing toward recycling and conserving.

My dad says that we are also helping the environment because we are recycling metal that would just go to the dump. My dad says that the metal we take to the scrap yard gets melted down and made into new things like bikes and toys. If we didn't scrap the metal it would just be lost forever in the dump.

*(Kevin)*

Keith and Kevin continued the meeting by showing students how the magnet would stick to the broken lamppost but not the brass apple. I had put a small pile of the reclaimed materials students had collected in the middle of the circle for the class to sort through. By the end of the morning, Keith had helped the class make a plan on how to sort their materials.

So what I do when I'm digging through the trash is I just quickly feel it to see if I think it is metal. One good way is to just tap it. If I think it is metal I throw it in the truck. You have to be real quick because there is so much trash to go through. Then after I fill my truck I take it all home and that is when I get my magnet out. I see what sticks and what don't and make two piles. You all can start by sorting your trash into what you think is metal and what you think is not metal. Then use this magnet I'm gonna give you to see which metal sticks and which metal don't.

*(Keith)*

The rest of the week, students sorted all of our reclaimed materials into the two piles. The following week, the students used the magnet to decide whether the metal was a ferrous or non-ferrous metal. Again, they sorted the pile of metal into two more piles. The students noticed that all of the aluminum cans we collected did not stick to the magnet. It was in these following weeks an unsettling consciousness began to rise within the students.

### Sorting trash in the classroom: the uncomfortable truth of being a trash producer

Several weeks after our initial conversations, students began having more serious conversations about their own impact on the earth. As we were going through our own recycled trash, one child commented,

Mr. Shaw, remember when Kevin's dad was in here and he taught us about going through trash and picking out the metal? He said that if he doesn't get the metal before the trash guys come, it just gets thrown in the dump and buried. Trash pollutes the earth.

Keith's visit heightened the value of recycling, at the forefront of our classroom meetings, this important realization became an ongoing part of the kindergarteners' conscientiousness. During the two-week-long project, the students began to realize the choices of what they placed in the trash can have real life consequences. With Keith explaining how he went around the city digging through the trash that everyone puts out on the curb and students bringing in items that they too would have just put in the trash, the idea was solidified. The abstract concept of where trash really goes became real. Students became aware that the trash they throw away does not actually just disappear. It gets buried in the earth and that these dumps, or places where trash ends up, take up space and can pollute our environment.

During our classroom meetings the decision was made to make an interactive bulletin board outside of our classroom. Here, other students of Patton were able to interact with the magnet and trash materials we had collected. We made signs sharing information we learned from Keith during his lesson about ferrous and non-ferrous metals, and also about what really happens to the trash we create and throw away. Our classroom became a popular stop for many students on their way to class in the morning or to the lunch room down the hall. Most mornings I found Kevin outside the door explaining to everyone that his dad was a scrapper and that his dad taught everyone in our class about ferrous and non-ferrous metals. Other students in our class would stand out in the hall with Kevin and talk about how recycling and reusing materials worked. Students explained how we were collecting materials to build our polar bear sculpture and how artists downtown did the same thing with their sculptures. The students from our classroom had conversations with their peers in the hallways about the importance of recycling and reducing the trash they produced.

Though learning about ferrous and non-ferrous metals was not a part of the district mandated curriculum, Keith's knowledge base—gained from his life experience—provided a deeply interesting and motivating learning opportunity for the students. This in turn helped students to focus on a stewardship towards the Earth through mindfulness of what they threw away in the trash and what they could recycle or reuse. Recycling and reusing are not main pillars of state or federal mandated curriculum in the classroom, although they should be in our current circumstances.

## Conclusion

By the end of the year the students and Mat the artist had created an 8-foot-tall polar bear sculpture from reclaimed materials (see Figure 7.2). The polar bear was standing on its hind legs on top of a mound of flattened aluminum beverage cans and other recyclable materials. Several other projects came from our working with Mat and trash. To continue the work throughout the school year I had to continuously renegotiate district mandated curriculum by closely tying the projects to kindergarten standards in mathematics, literacy, and language. The trash and recycled materials became substitutes for math curriculum manipulatives. We read books about trash and art, we created with trash during literacy lessons, but I was bound to carefully construct literacy and language standards to make book selections.

In this chapter I have chosen to use the description of social and financial clout as a means of critical reflection towards work in the classroom. Theorists have argued that the common public school classroom is a tool of systemic societal reproduction (Bourdieu & Passeron, 1979; Freire et al., 2014; Hyun, 2006; Lareau, 2000; Rishel, 2008). Like institutionalized cultural capital, clout plays a large role in whose voice is heard in curricular decision making, not only locally in the classroom but also in a broader political perspective of educational policy making. Connecting clout to Freire's view of historic fatalism, the

**FIGURE 7.2**  "Our polar bear sculpture"
*Source*: Jonathan Shaw

market, and dehumanization of American public education, I used the tool of negotiated curriculum to enact localized change within the larger societal production of American public education. My hopes are that Kevin and his dad Keith saw themselves as active co-constructors of our public-school culture and that my students saw themselves as larger members of the local Patton community, downtown, and beyond. To disrupt what I perceived as a narrow set of educational goals and standards largely focusing on literacy and mathematics in the kindergarten public classroom was not my original intention, but putting trash, the arts, Keith, and Kevin in the center of the room allowed this to happen. I make this argument because, like many teachers in North America, a large portion of my job performance evaluation was tied directly to my students' ability to perform on district literacy and mathematics assessments. Teachers in low-income schools whose students do not make adequate yearly progress—as determined by standardized tests that are designed with the norms of white, middle-class students in mind—are in danger of losing their positions, no matter how engaging, open-minded, or creative they might be to students and families. This consequence is a direct result of state and federal educational policy and funding guide lines that narrow curriculum. Students' standardized test scores in literacy and mathematics in the US today are used to evaluate a school's performance and if schools do not meet specific performance indicators a school could be "taken over" by state officials in a move to improve literacy and math test scores. Unfortunately, such moves also frighten teachers into narrowly defined curriculum choices that limit their students' capacities and joy for learning, and prevent parents like Keith from having a say in learning, effectively shutting down organic opportunities to connect with the specialized knowledge of elders in marginalized communities.

Were it not for a pivotal moment at the parent–teacher conference, our classwork would have been business as usual. I had all the school district's prescribed focus of test scores sitting on the table as Keith walked in with his scraps of metal. As a middle-class, educated white male working for the school district, I held all of the conventional power in the relationship between family and school that critical educators attempt to disrupt (Kroeger & Lash, 2011). As a new teacher, I knew what was supposed to happen that night. As Keith set his items on the table, I pushed aside the test scores, the school district's agenda and removed the overlying dominant social clout from the parent–teacher interaction. In doing so, an organic, earth conscious educational experience grew. In true serendipitous fashion, as I wrote this chapter, I received an informational pamphlet in the mail about my community recycling program which included a list of items community members were allowed to put into our recycling receptacles. I was shocked to discover that glass bottles were no longer being accepted—glass bottles are no longer a profitable recyclable material according to the local recycling company. It is unfortunate that such an important and impactful environmental decision is solely relegated to financial and profit based constraints. At the conclusion of this chapter, and as a more experienced teacher than I was during my time with Keith and Kevin, I still have many questions.

Does our current place in American history allow for educators to navigate politics and economics to be able to incorporate more Earth-minded lessons into the curriculum? The optimist inside me says yes; the pragmatic within me says no. I say no because I realize that as individuals we work within a larger societal system; no matter how critically and conscientiously we become aware of economic constraints, the system still exists. As an individual, I had a choice to make in the classroom to either maintain a status quo with my educational practice, or break free and work against the expectations. American society has a choice as well. We have established an educational system in the United States that promotes consumption, production, and upward economic mobility as the main values to be taught; "the invisible hand of the market" further supports an underlying culture of capitalistic values (Freire et al., 2014). Our congress has placed more value on mathematical and literacy education by funding research in the public sector and universities to further work in these fields, neglecting citizenship, art, and environmental science. When research beyond math and literacy are funded—for such purposes as the understanding and prevention of climate change or towards eco-focused curriculum—local communities will need to place higher value on environmentally sound practices over profitability. Educators will have to work against the larger system of public interests in capitalism in the United States.

## References

Bourdieu, P. (1986). The forms of capital. In J. Richardson (Ed.), *Handbook of theory and research for the sociology of education* (pp. 46–58). New York: Greenwood.

Bourdieu, P., & Passeron, J. C. (1979). *Reproduction in education, society and culture*. Beverly Hills, CA: Sage.

Doucet, F. (2008). How African American parents understand their and teachers' roles in children's schooling and what this means for preparing preservice teachers. *Journal of Early Childhood Teacher Education, 29*(2), 108–139.

Freire, P., Araujo Freire, A., & De Oliveria, W. (2014). *Pedagogy of solidarity*. Walnut Creek, CA: Left Coast Press.

Hyun, E. (2006). *Teachable moments: Re-conceptualizing curricula understandings. Studies in the postmodern theory of education*. New York: Peter Lang.

Kroeger, J., & Lash, M. (2011). Asking and listening and learning: Toward a more thorough method of inquiry in home-school relations. *Teaching and Teacher Education, 27*, 268–277.

Kuby, C. R. (2013). *Critical literacy in the early childhood classroom: Unpacking histories, unlearning privilege.* New York: Teachers College Press.

Lareau, A. (2000). *Home advantage: Social class and parental intervention in elementary education.* Lanham, MD: Rowman & Littlefield.

Lareau, A., & Horvat, E. M. (1999). Moments of social inclusion and exclusion: Race, class, and cultural capital in family–school relationships. *Sociology of Education, 72,* 37–53.

National Governors Association Center for Best Practices, Council of Chief State School Officers (2010). *Common core state standards.* Washington, DC: National Governors Association Center for Best Practices, Council of Chief State School Officers.

Rishel, T. (2008). From the principal's desk: Making the school environment more inclusive. In T. Turner-Vorbeck& M. Miller Marsh(Eds.), *Other kinds of families: Embracing diversity in schools* (pp. 46–63). New York: Teachers' College Press.

# 8

# EATING FOR ECOLITERACY

## The social praxis of sustainability at a residential environmental education center

*Kate Albing*

## The big crooked river and the watershed

From aerial photographs, Cuyahoga Valley National Park (CVNP) registers as a sprawling patch of green lodged between the concrete grids of Cleveland (to the north) and Akron (to the south). The Cuyahoga Valley Environmental Education Center (CVEEC) sits in the southwest corner of the Cuyahoga Valley National Park, and constitutes 500 acres of the park's 33,000 acres. From August through June, schools from urban, suburban, and rural areas of northeastern Ohio and beyond explore the campus as participants in our residential environmental education program, All the Rivers Run. For four days and three nights, the children live in the CVEEC's dorms, eat in its dining halls, and hike past its meadows, wetlands, ponds, forests, and creeks.

As a field instructor at the CVEEC, I pose to children the question, "What does it mean to live in a healthy watershed?" I then attempt to elicit answers to this inquiry through a variety of interconnected experiences. The children and I travel by bus to the banks of the Cuyahoga River and become citizen scientists, testing the chemical and physical properties of the river water. We seek out time capsules hidden along the trail that contain artifacts of the watershed's prehistoric past (fossils and coal), its industrial past (a railroad spike and rubber tire), and its agricultural past (a horseshoe). We sing songs, play games, and tell stories that emphasize the role of the watershed and its implications for the health of all living things.

We bring an understanding of the watershed to the forefront of our learning at the CVEEC for a number of reasons. First, it allows us to launch into other ecological concepts, such as the foundational importance of water to all life on Earth. Second, it offers the children a new lens through which to interpret their surroundings: we live within the Cuyahoga River watershed, and this means that the headwater creeks on our campus flow into the Cuyahoga River. The Cuyahoga River empties into Lake Erie, one of the five Great Lakes, and this water finds its way over Niagara Falls—a recognizable touchpoint for most children—and out the St. Lawrence Seaway to the Atlantic Ocean. By tracing our river's path, we illustrate our connection to a global system. And finally, the watershed brings attention to the fact that our park wouldn't exist if it weren't for the Cuyahoga River, and its use—and profound misuse—over time.

## The burning river

The Cuyahoga Valley National Park was established in 1974 partly in response to the eco-logical crisis precipitated by industrial practices of the time. Factories effectively treated the Cuyahoga River as a dumping station for industrial waste. The region has supported industries such as paper mills, freight hauling via trains, mining, rubber, tire, and steel plants, farming, automotive and machine factories, sawmills, and at times even quarries, all of which take a tre-mendous toll on the health of the watershed. The Cuyahoga River received national media coverage for the frequency with which it caught fire (13 times) due to the layers of oil, sludge, sewage, and debris floating on its surface between the years of 1868 and 1969. It was the fire of 1969, though, that attracted national attention and contributed to legislation for the Great Lakes Water Quality Agreement and the Clean Water Act of the 1970s. The river has not caught fire since its infamous 1969 ignition.

Although healthier today than 40 years ago, the river is still impacted by human activity. The lower 46.5 miles of the Cuyahoga River remain one of 43 Great Lakes Areas of Concern—waters in the US and Canada that have experienced environmental degrad-ation, fail to meet the objectives of the US–Canada Great Lakes Water Quality Agree-ment, and are impaired in their ability to support aquatic life or beneficial uses (Ohio Environmental Protection Agency, 2014). Since the Center's establishment in 1994, field instructors at the CVEEC have conducted their teaching against this backdrop of ecological disaster and ongoing mitigation.

Arguably, the treatment of the Cuyahoga River stemmed in part from a perceived detachment of people from their environment. David Orr (2013) wrote that "to a great extent, we are a deplaced people for whom our immediate places are no longer sources of food, water, livelihood, energy, materials, friends, recreation, or sacred inspiration" (p. 2013). At the CVEEC, my fellow field instructors and I attempt to counteract this sense of "depla-cement" amongst children by connecting them to the origins of the resources that sustain all life on earth.

## Residential environmental education in a national park

Environmental education has always been at the heart of Cuyahoga Valley National Park's mission. The CVNP has long recognized the value of residential programs in particular. The CVNP's *Plan for environmental education* (United States Department of the Interior, 1986) references the 1977 General Management Plan as prescribing a facility to be "adapted for use as a major environmental education center, with classrooms, labs, a nature center, and dormitory space. The surrounding lands will be used for nature study and research" (p. 7). The CVNP saw that education can be wielded as a tool for resource management: "because the roots of the environmental problem may be matters of ignorance, values, and attitudes [more] than anything else, education holds the greatest long-range promise of solution. In this regard, environmental education is in everyone's best interest" (p. 1). The CVEEC is the realization of the park's vision for an environmental education program that reorients young people's attitudes toward the outdoors.

It is worth noting that, unlike many other national parks, the CVNP is situated between two major cities, Cleveland and Akron. Targeting urban populations is a key part of the park's and the CVEEC's missions. However, this goal requires the recognition that many people in these urban centers have significant physical, cultural, and emotional barriers to connecting with green spaces. A personal example illustrates this reality: in 2017, fifth grade students from Willson Elementary School in Cleveland, Ohio stayed at the CVEEC. In preparation for their visit, I met with the students' teachers to discuss ways to support the students, many of whom are deaf and hard of hearing. One of the teachers told me that, despite the neighborhood's proximity to Lake Erie (1.8 miles away), she believed that the majority of her students had never seen its waters. Their conception of the outdoors, she told us, is limited to what they glimpse out of their school bus window on the route to and from school. We know that many of the children from urban settings who visit our school have fraught relationships to the outdoors. We address this reality in a number of ways: staff receive training in responding to the needs of diverse populations, donors funded scholarships for more than 3,000 youth to attend programs at the CVEEC in 2017 alone, and summer programs shuttle children for free from Cleveland and Akron to our rural park, which, as of this writing, remains largely inaccessible via public transportation.

## Mapping the terrain of environmental education

New frameworks for environmental education have emerged in the past three decades. In the mid 1990s David Orr and Fritjof Capra coined the term "ecoliteracy" (Capra, 1995; Orr, 1992) to describe humanity's ability to understand the basic principles of ecology and live accordingly. Because of its applicability to both social and ecological systems, it is this new educational paradigm that I have found most relevant in informing my approach to, and the evaluation of, residential environmental education and its social dimensions in particular.

In its *Plan for Environmental Education*, the Division of Interpretation and Visitor Services defines environmental education as a holistic endeavor, and lists a number of "definitions" in an attempt to map out all the possible implications of a complex, interdisciplinary field. The document contains no guidelines for the design or facilitation of meals on campus, except to say that groups who stay would be responsible for their own meals and washing their own dishes. The absence of an official policy for this part of the program opened spaces for me—and for the field instructors who came before me—to use the endlessly engaging practice of mealtime to reimagine the scope of children's environmental education to include social systems.

## Living systems principles

Living systems frameworks provide a lens through which to view the curriculum components of the CVEEC. Linda Booth-Sweeney (2009) defined living systems in simple terms:

> We use the phrase living systems as a metaphor, to represent an animate arrangement of parts and processes that continually affect one another over time. There are living systems on all scales, from the smallest plankton to the human body to the planet as a whole.
>
> *(p. 3)*

In *Ecoliteracy: Mapping the Terrain* (2000), Capra identified six principles that represent the interactions that occur in all living systems. These principles must be present for a system to be considered sustainable. In Table 8.1 I list and define these principles and relate them to the processes that unfold in/around Furnace Run, a Cuyahoga River tributary that students visit during their stay at the CVEEC.

During our closing ceremony on the final day of the four-day program, we ask children to tell us their favorite thing about their time with us. They inevitably and resoundingly

**TABLE 8.1** Tapping into the transformative potential of mealtime

| Living systems principle | Definition (adapted from Capra, 2000) | Curricular example |
| --- | --- | --- |
| Networks | All living things in an ecosystem are interconnected through networks of relationships. They depend on these networks to survive. | The cottonwood trees bordering Furnace Run creek depend on the shade of neighboring trees to keep temperatures low and humidity high—they are better able to survive in a forest than in isolation. |
| Nested systems | Nature is made up of systems that are nested within systems. Each individual system is an integrated whole and—at the same time—part of larger systems. Changes within a system can affect the sustainability of the systems that are nested within it as well as the larger systems in which it exists. | The land that drains into Furnace Run is that creek's watershed, but that watershed is itself just one small component of a larger watershed. |
| Cycles | Members of an ecological community depend on the exchange of resources in continual cycles. Cycles within an ecosystem intersect with larger regional and global cycles. | Evaporation from the surface of Furnace Run and transpiration from the trees condense into clouds, which rain back onto the land that drains into Furnace Run. |
| Flow | Each organism needs a continual flow of energy to stay alive. The constant flow of energy from the sun to Earth sustains life and drives most ecological cycles. | The dead leaves from the cottonwood trees feed the crayfish in Furnace Run, which feed the muskrats, whose droppings fertilize the soil. |
| Development | All life—from individual organisms to species to ecosystems—changes over time. Individuals develop and learn, species adapt and evolve, and organisms in ecosystems coevolve. | The post-industrial and post-agricultural landscape surrounding Furnace Run has developed into a forest over the past 40 years. |
| Dynamic balance | Ecological communities act as feedback loops, so that the community maintains a relatively steady state that also has continual fluctuations. This dynamic balance provides resiliency in the face of ecosystem change. | The invasive emerald ash borer has decimated the ash trees in the forests around Furnace Run. Over time, different trees will repopulate and redefine the landscape. |

declare "the food" as one of the most positive parts of their experience. I set out to explore the origins of these highly positive associations around mealtime. Close examination revealed that the structure and practices we implement in the dining hall closely align with the living systems principles. One of the implications of this finding is that social experiences that embody living systems principles can be a site for transformation, eliciting new and exciting understandings of oneself, one's relationship to others, and one's relationship to the earth.

The children on our campus eat a total of ten meals, which amount to ten hours, in our dining hall. Rather than functioning as a "break" from our environmental education curriculum (see Figures 8.1 and 8.2), those 10 hours of mealtime have come to serve as a critical opportunity to reinforce an awareness of nature's patterns and processes though a somatic experience.

**FIGURE 8.1** Dining, Lipscomb Hall
*Source*: Kate Albing

**FIGURE 8.2** Exterior, Lipscomb Barn
*Source*: Kate Albing

## Key aspects of living systems

Below, I discuss two important characteristics common to all living systems, and their implication for the structure of eating experiences: living systems are lively systems, and living systems are relational.

### Living systems are lively systems

Living systems awareness requires somatic, embodied experiences. Widhalm (2011) wrote, "From a living systems awareness perspective, life is a vibrantly felt experience. Our bodies are our primary access to the felt, visceral experience of life" (p. 46). What could be a more appropriate setting for fostering living systems awareness than the dining room table, a site where we energize and revitalize our bodies? (Figure 8.1) Indeed, meals are a time not simply for physical nourishment, but for the nourishment of our hearts and minds as well. Author and activist Michael Pollan (2014) agreed: "the shared meal is no small thing. It is a foundation of family life, the place where our children learn the art of conversation and acquire the habits of civilization: sharing, listening, taking turns, navigating differences, arguing without offending" (p. 8).

As a highly sensorial and social experience, eating in community has the capacity to be a joyful, nourishing endeavor that evokes nature's capacity for sustaining life. Stone (2009) posited that any sustainable community worth striving toward is alive in every sense of the word. At the Cuyahoga Valley Environmental Education Center, field instructors envision mealtime as a highly lively ritual. A strand of joy, dynamism, and celebration runs through each practice we implement in our efforts to cultivate living systems awareness.

### Living systems are relational

Living systems awareness is an invitation to reconsider the world and one's place in it. Capra (2013) explained that "when you look at the principles of ecology in detail—networks, diversity, cycles, partnership … you can interpret them also as principles of community. Ecology and community go hand in hand" (p. 216). Through the development of living systems awareness, we come to see that, ontologically, there is no sense of person without a sense of the community. With this relational ontology in mind, we center our dining hall design, practices, and intentional awareness on nurturing relationships.

## Overview of eating practices in American public schools

It may be helpful to offer some perspective on the eating experiences of most American school children. The School Nutrition Association's (SNA) *State of School Nutrition 2018* survey, which included responses from nearly 1,000 SNA member school districts nationwide, found that the typical lunch period length for elementary schools is 25 minutes. This includes the time needed to get to and from the cafeteria and wait in line to get food. Approximately one-third of all food is wasted at the retail and consumer levels. According to the United States Department of Agriculture (USDA) (2016), food wasted by children is similar to the rest of the U.S. population.

Potentially exacerbating the problem of waste is the fact that many children do not choose the quantity of food they receive. Rather, they are issued a federally mandated meal. According to the USDA, more than 30 million children received free or low-cost lunches in school cafeterias across the United States in 2016. Beginning in the 2012 school year, the USDA required that students take both a fruit and a vegetable with their lunch (2016). A recent study (Just & Price, 2013) suggests that requiring students to take fruits and veggies increases waste substantially.

## Living systems principles at the Cuyahoga Valley Environmental Education Center dining hall

Eating practices at the CVEEC contrast starkly with the practices in place at many schools. Rather than base our eating practices on convenience, efficiency, and federally mandated nutritional guidelines, we instead attune our practices to sustainability. This approach aligns with the work of Fritjof Capra (2013), who suggested that "we do not need to invent sustainable human communities from scratch but can model them after nature's ecosystems because nature's ecosystems are, in fact, sustainable communities" (p. 202). Below, I reiterate the six living systems principles and describe how they manifest in our community-based eating practices.

### Flow: shake the hand that feeds you

To initiate the children's awareness of flow, we introduce them to the people who prepare their food. Unlike many dining facilities, the CVEEC dining hall features a wide window that invites students to observe the kitchen staff as they prepare meals and allows them to ask questions about ingredients, recipes, and techniques. At the children's first breakfast, we display a poster with photos of the kitchen staff, and share with the children the names of the chefs, the

**FIGURE 8.3**  Hoop house
*Source*: Kate Albing

number of years they've worked at the CVEEC, and a fun fact about them. Chef Jim used to be an art teacher, for instance, and Chef Melissa's favorite superhero is Batman. This provides an entry point into conversation and connection. Notably, the chefs have a unique relationship to much of the produce they prepare: they grow it from seed. Behind the dining hall, a 500-square-foot hoop house contains eight raised beds in which the kitchen staff cultivates a variety of fruits, vegetables, and herbs (see Figure 8.3). As part of their campus orientation, children may accompany one of the dining hall staff on a tour of the hoop house. They also see our compost piles, the final destination for their uneaten food. Here, they are invited into a sensorial experience of energy flow: they smell the compost's earthy scent, feel the heat produced by the aerobic breakdown of nutrients, and see the orange peels, dinner rolls, and napkins from their prior meals in various stages of decomposition. This compost pile, the staff explains, will be used to nourish the plants growing in the hoop house. By connecting the students with the origins of their meals, the students gain a deeper understanding of the energy required to produce them (see Figure 8.4).

## Networks: making new connections

When the students first enter the dining hall, the instructor responsible for meal facilitation assigns the learners randomly to one of seven large round tables that seat eight people. This act aims to disrupt existing social dynamics, and create the conditions under which new connections can form. CVEEC instructors spread out amongst the tables and eat their meals with the students. They set an intention to get to know each child by facilitating conversations and connections. They tell jokes, ask questions about the children's lives

**FIGURE 8.4** Compost bins
*Source*: Kate Albing

outside school, and play games. For many students, this arrangement represents a radical departure from the conventional student–teacher dynamic. The act of eating with students creates opportunities for heartfelt connections: the table acts as a unifier, a place of community. Children and adults reflect on the shared experience of the day's events, they compare their observations about the frog in the pond or the spider in the hollow tree, they swap stories and recall memories. This shared time around the meal table can facilitate the formation of a vibrant, energized, and multi-generational learning community.

## Dynamic balance: weighing waste

Once the meal ends, the meal facilitator asks the children to consolidate all the uneaten foods, liquids, and compostables at their table. One person per table then deposits the refuse in their respective color-coded buckets. The meal facilitator weighs the aggregated liquid and solid waste in front of the children, writes the results on a large whiteboard and announces those results. The facilitator asks the children, "What happened during this meal to influence the amount of food that ended up in those buckets?" Children may respond that they took more than they could eat, or that they were particularly hungry because of that afternoon's hike, or they had never eaten the from-scratch dishes we prepare, and thus weren't ready for the unexpected flavors. This question invites the children to reflect on the way their individual choices impacted the overall outcome of the group's food and liquid waste. The problem of food waste in the United States may be overwhelming—a Natural Resources Defense Council (Grunder, 2012) report found that 40% of all the food produced in the U.S. is thrown away—but the lived experience of addressing food waste, of observing the way each small individual change can manifest as a significant collective change, can help the students identity their own agency within community.

## Nested systems: the round table

The children in our dining hall sit at the same round tables throughout the duration of the week. At these tables, each individual nourishes his or her body—its own nested system of cells and organs—aware that the others at his or her table do the same. It is the job of each individual at the table to monitor the amount of food and liquid he or she consumes. The meal lasts nearly an hour, far longer than the duration of mealtimes at most schools. We suggest that children eat everything on their plate before they put more food on it. In this way, they are invited to develop self-regulation, and hone in on the sensations of hunger and satiety. At the end of each meal, all the individuals at the table consolidate their table's unfinished food and liquid. This process allows each individual to put their choices into the larger context of their table. And finally, at the end of the meal, every table brings up their table's waste, combining it with the waste of every other table. This lets the children observe how their individual choices reverberate throughout the system of their table, which then reverberates through the system of the dining hall as a whole.

## Cycles: Meal as ritual

Meals in our dining hall are highly structured processes. Each meal begins with a facilitator calling the learners to attention with "ago," a word from the Twi language of West Africa that means "I am asking for your attention and respect." Learners reply in unison with "ame," a word that signifies their willingness to give their attention and respect. Once the dining hall is silent, the meal facilitator reads the written menu from a centrally displayed whiteboard. The menu includes serving sizes for each item; it is only after everyone in the dining hall has had the opportunity to receive this serving size that anyone can come up to the buffet table to get another serving. The facilitator then reads to the children an eco-logically oriented quote such as, "What we do to the earth we do to ourselves" (attributed to Chief Seattle). A moment of silence is observed to reflect on the quote, and then tables are called up one by one to serve themselves from the buffet. Every meal ends with the highly structured waste consolidation and weighing process, as outlined above. Finally, once all the waste has been weighed and learners have discussed the results, everyone in the dining hall—adults and children alike—rise from their tables for a facilitator-led "digestive song," characterized by repetition, movement, and wordplay. All adults in the dining room—teachers, field instructors, even, occasionally, kitchen staff—participate. It is a joyful, celebratory closing to the meal.

It may not be immediately apparent how these rituals contribute to living systems aware-ness. However, it is these powerful, lived experiences that catalyze social transformations. Widhalm (2011) agreed, stating that "there are strong synchronicities between living sys-tems principles and the design components of ritual" due to their power to foster trans-formative experiences. Takahashi (2004) echoed this sentiment in his writing about the power of art for social transformation:

> Many of the nonformal educators/activists, in spite of the hardships they were facing, had the amazing knack of celebrating the joy of life with their whole beings. When I joined them in singing, dancing, and laughing, I felt whole and alive, and I learned the joy of living.
>
> *(p. 175)*

## Development: Change from within

Our meal structure seeks to counter the narrative of modern industrial society, primarily the view of nature's sole value as a resource to be exploited. Depending on the age of the children residing on our campus, we might introduce statistics and data about the tremen-dous environmental problem of food waste. Globally, the Food and Agriculture Organiza-tion of the United Nations estimated in 2011 that one-third of all the food grown is lost or wasted, an amount valued at nearly $3 trillion. However, statistics such as these are nearly unfathomable to most young learners. As Takahashi (2004) wrote:

> The emphasis on facts about the symptoms of environmental destruction can give students the notion that environmental issues are something that are "out there" … this approach has a disempowering effect on students by giving them the sense of power-lessness and despair.
>
> *(p. 170)*

How then, do we motivate children to act in sustainable ways in relation to the food they eat? How do we spark this shift in consciousness? We do this in our dining hall by empha-sizing the agency of each child to do a small part to lower the amount of food and liquid waste. In simple words: we teach the children that their food matters, that the people who prepared their food matter, and that they matter. With every subsequent meal, learners' relationships to their food changes in qualitative and quantitative ways. Learners may come to see differently the crusts on their sandwiches, the slightly green slice of cantaloupe that, before, they may not have thought twice about throwing away prior to their time with us. And because we write the amount of food and waste on our large, prominent whiteboard, the children are able to keep track of the development of their relationship to food; they actually see the numbers diminish or grow throughout the week. Rather than shame or guilt the children for their waste, we celebrate their success and ask them to consider the energy poured into their food's production, and the implications of throwing it away.

## Conclusion

Shared meals hold tremendous potential as sites for radical social transformation. By structuring meal practices in accordance with living systems principles, we can positively impact children's relationships to their bodies, to each other, and to the earth itself.

How might schools look and feel if living systems awareness determined the social practices of mealtimes? Below, I list several questions for teachers who wish to explore how living systems principles might translate into concrete pedagogical action.

- In what ways can we foster connections between students and the origins of their food? Can we introduce kitchen staff, display their photos, or express gratitude to staff through a letter, poem, or song?
- Can we make time to sit and eat with our students?
- How is the meal structured? Is it possible to incorporate ritual (song, poetry, intention) into the mealtime?
- How can we foster the formation of new friendships and feelings of belonging? Should we include conversation starter cards at each table or regularly mix up seating arrangements?
- Are children cognizant of where their uneaten food goes? What can teachers do to foster such an awareness? Can we weigh the food, post photos of a landfill or recycling plant on respective disposal bins, or visit a recycling plant?
- Are there unused physical spaces inside and outside the building in which children can begin to maintain a small garden or compost their food scraps?
- What is the physical space in which children dine? Are they able to see and speak to one another? Do they sit with the same people every time they eat?

## References

Booth-Sweeney, L. (2009). *Connected wisdom: Living stories about living systems.* SEED.

Capra, F. (1995). *The web of life: A new scientific understanding of living systems.* New York: Anchor Books.

Capra, F. (2013). Deep ecology: Educational possibilities for the twenty-first century. *The NAMTA Journal, 38*(1), 201–216.

Capra, F. (2000). In M. Crabtree (Ed.), *Ecoliteracy: Mapping the Terrain.* Berkeley, CA: Learning in the Real World.

Food and Agriculture Organization of the United Nations. (2011). *Global food losses and food waste: Extent, causes and prevention.* Rome: FAO.

Grunder, D. (2012, August). *Wasted: How America is losing up to 40 percent of its food from farm to fork to landfill.* Natural Resources Defense Council Issue Paper IP: 12-06-B.

Just, D., & Price, J. (2013). Default options, incentives and food choices: Evidence from elementary-school children. *Public Health Nutrition, 16*(12), 2281–2288.

Ohio Environmental Protection Agency. (2014). *Delisting guidance and restoration targets for Ohio areas of concern, Version 3.*

Orr, D. (1992). *Ecological literacy: Education and transition to a postmodern world.* Albany, NY: State University of New York Press.

Orr, D. (2013). Place and pedagogy. *The NAMTA Journal, 38*(1), 183–188.

Pollan, M. (2014). *Cooked: A natural history of transformation.* New York: Penguin Books.

School Nutrition Association. (2018). *School nutrition operations report: The state of school nutrition 2018.* Retrieved from https://www.fns.usda.gov/school-meals/creative-solutions-ending-school-food-waste

Stone, M. K. (2009). *Smart by nature: Schooling for sustainability.* Healdsburg, CA: Watershed Media.

Takahashi, Y. (2004). Personal and social transformation: A complementary process toward ecological consciousness. In E. O'Sullivan (Ed.), *Learning toward an ecological consciousness: Selected transformative practices* (pp. 169–182). New York: Palgrave Macmillan.

United State Department of Agriculture. (2016). *HHFKA implementation research brief: Fruits and vegetables.* Retrieved from https://fns-prod.azureedge.net/sites/default/files/ops/HHFKA-FruitsVegetables.pdf

United States Department of the Interior. (1986). *A plan for environmental education for the Cuyahoga Valley National Recreation Area.*

Widhalm, B. (2011). *Nature as guide for vibrant learning: A living systems framework for academic learning experience design toward a thriving sustainable world.* (Doctoral dissertation).

# PART III

# Sustainable futures

## New terrestrial collectives

# 9

# NATURE CAN BE DEAD AND ALIVE

## Pachysandra is a bad guy

*Adonia Porto*

## Introduction

Our sole existence as a living species naturally connects us with the living earth (Kahn & Kellert, 2002; Sobel, 2008). "Naturally" is a term referring to the force that drives us to exist in a certain way such as our instinct to do something as well as "being born of the earth" or being "earth's child" (Corsaro, 1997; Demeritt, 2001; Sobel, 2008). This natural force connects us directly to nature if we spend enough time in nature for the connection to exist. Our connection reminds us that the natural world is alive and conscious and needs our care (Chawla, 1998, 2007). Additionally, it forces our recognition that we are like all living things, needing essential elements of earth to sustain our lives. Some adults maintain their connection from childhood upholding the need for earth's protection, while others lose the connection and use earth for personal gains (Demeritt, 2001). Authors note that if children develop a strong enough connection as children, it will last into adulthood (Chawla, 1998, 2007; Sobel, 2008).

Leading into a research study with children, I grappled with how I defined nature as an adult and how children might define nature in the study. Prior to our work, I defined Nature as:

> the grass, the trees, and the things most associated with green. Nature is a place to go· and a space to be in. When fully immersed, people become nature. They develop a strong connection where they realize, or perhaps remember, they are nature them-selves. ... I define nature as how we exist today. We survive in nature. We use its resources to sustain our lives but we protect it too. All things considered, nature is the future as it will exist long after our time has ended. Nature is timeless.
>
> *(Journal entry, August 14, 2016)*

This definition inspired me to remember being a child in the grass and imagined that children in the study might describe nature as the things outside their school and homes. I assumed it may be expressed as the earth, the soil, the grass, the trees, the water, the sky, the sun, and the moon. Nature may be their families, babies, puppies, and the idea of being loved. My assumptions of how children might talk about nature during the study helped me consider what questions to pose and how to listen for opportunities to understand and generate meanings with children.

## Supporting children's experiences in nature

Dependent on government and state policies, teachers in North American classrooms are bound to meet state and national standards which leave little room for implementation or funding for desired equipment, materials, and experiences that support children's needs and interests in the outdoors (Dyment & O'Connell, 2013; Little & Eager, 2012; Little & Wyver, 2008; Stan & Humberstone, 2011). Also, teachers feel less compelled to advocate for children's experiences in the outdoors if their own experiences have been less than favorable (Kellert, 2002; Sobel, 2008). Some teachers feel as though they do not have proper training to be effective teaching outdoors, or prefer to stay indoors to avoid inclement weather and getting dirty (Copeland et al., 2012).

Regardless of existing barriers, two priorities are necessary for meaningful curriculum in the outdoors. Children must experience uninterrupted time in natural environments outside of their primary playground fence. Exposure to diverse landscapes provides children endless opportunities for creative play and inventiveness of their own actions, unlike a fenced in area regulated by adults (Fjortoft, 2001, 2004; Fjortoft & Sageie, 2000; Staempfli, 2009). When possible, adults should observe how children relate and connect with(in) nature from a distance but join in conversation with children when invited or needed to ensure safety. Kernan and Devine (2010) confirmed, "the most prevalent value attributed to the outdoors was freedom" and moving freely was considered natural and "a necessary part of being a child" (p. 377). This freedom provides opportunities for independence and self-esteem (Murray & O'Brien, 2005). If children are given enough time to build a relationship with nature in a context with adult support, children will find value in caring for nature and understand how their own actions affect the environment (Murray & O'Brien, 2005; O'Brien & Murray, 2006).

The second priority is for teachers to be mindful of their relationships with children, where they trust children to approach risk-taking in the outdoors while providing them with agency to make their own responsible decisions about preferences for being with(in) nature. Stifling children's risk-taking opportunities to reduce adult stress about safety often lead to more unsafe risks as children feel too managed by adults (Little, Wyver, & Gibson, 2011; Stan & Humberstone, 2011). Furthermore, teachers must keep in mind that their attitudes, if negative, influence children's values of nature and natural environments (Kahn & Kellert, 2002; Sobel, 2008).

## Listening as a context for research with children

With these priorities in mind, I considered how additional classroom practices would impact our time together in nature. My goal was to uncover children's experiences with (in) nature while providing them with as much uninterrupted time as possible. I purposefully use the term *with(in) nature* as a theoretical move in this chapter to suggest the complexities of children's understandings as they experienced nature as something that was a part of them, recognizing they were nature themselves, but also as a space to be in outside their typical classroom environment. I also use the term *(re)know* to suggest how each time we experienced nature together we were always in a position of being, becoming, and (re) knowing (Davies, 2014). Practicing listening (Rinaldi, 2006) as a way of being in the classroom allowed me to step back and comfortably follow the leads of children and be a part of their constructions of nature as phenomena while only contributing to the conversation when invited or needed to ensure the comfort and safety of the children.

Listening goes beyond what is heard and refers to a pedagogy practiced by Reggio educators. Rinaldi (2006) suggests that listening should incorporate several parameters into the curriculum context but these five directly influence my own practice: (a) "listening as sensitivity to the patterns that connect, to that which connects us to others," (b) "listening as time, a time that is outside chronological time—a time full of silences, of long pauses, an interior time," (c) "listening is an active verb that involves interpretation, giving meaning to the message and value to those who offer it," (d) "listening that does not produce answers but formulates questions," and (e) "listening is not easy ... . [i]t requires a deep awareness and at the same time a suspension of our judgments and above all our prejudices" (pp. 80–81).

Listening, therefore, is "a listening context," in which one learns to listen and narrate, where individuals feel competent to represent their theories and offer their own interpretations of a particular question. In representing our theories in a community, we "re-know" them, making it possible for our images and intuitions to take shape and evolve through action, emotion, expressiveness, and iconic and symbolic representations (Rinaldi, 2006). Listening in my classroom is daily extended conversations with children that create an environment for building meaningful relationships where children share who they are (their culture), what they know (knowledge they have or knowledge they would like to have), and how they interpret the world (how they learn). Over time, learning in a listening context encourages children's competence to be leaders of the community and guide their own learning both in the classroom and outdoors. Listening to children in nature gave us freedom to escape the daily routine indoors and grasp nature for what it is and what it meant to us.

The Mosaic Approach (Clark & Moss, 2008) provided us with ways of reflecting together to create meanings, and allowed me to capture children's pre-reflective and reflective experiences while considering their right to choose when and how to participate as well as how much to "say" about their experiences in the wetland environment. The Mosaic Approach is a meaningful way research can be conducted with young children (Clark & Moss, 2008; MacNaughton, 2003; Rinaldi, 2006) and essentially asks, "What does it mean *to be you* in this place now, in this present moment, in the past, and in the future?" (Clark & Moss, 2008, p. 17). The methods adapted for this study were Magic Carpet and child-interviewing. Magic Carpet was revised to offer videos, as opposed to a slideshow of photographs, captured by the children or me, offering a less product-oriented level of engagement that was welcomed by the children who did not feel as comfortable drawing, writing, or taking photographs. Child-interview, a method of short semi-structured interviews, was also revised to be an open invitation for reflection that could occur at any moment initiated by the children or me.

## Phenomenology with children: methodology and background of study

The goal of phenomenology is to "come to a deeper understanding of the nature or meaning of our everyday experiences" (van Manen, 2014, p. 37). As a teacher, I wondered how studying nature with children as everyday experiences would lead to new understandings about nature's phenomena. As a researcher, I wondered how attempting to simplify the act of just being in nature would allow me to exist in a stance of being and becoming with children (Davies, 2014). This prompted me to investigate preschool children's pre-reflective and reflective experiences to discover the phenomenology of nature as both a space and something we are part of as humans. Twelve preschoolers, ages 4 and 5, experienced nature in a wetland environment near their school with myself and an outdoor educator over the course of nine months. Methods of listening (Rinaldi, 2006) partnered with adapted methods of The Mosaic Approach (Clark & Moss, 2008) and phenomenological writing of anecdotes with punctum (van Manen, 2014) allowed me to generate nature understandings with young children. Three of the 14 final anecdotes featured in my research are shared in this chapter to emphasize how children's construction of nature changes over time, along with what children considered significant about nature which may impact their views of nature later in life.

## The phenomenological experience

Max van Manen (2014) defined phenomenology as the project that tries to describe the pre-reflective meaning of the living now. However, phenomenology is also aware that when we try to capture the "now" of the living present in an oral or written description, then we are already too late. The moment becomes objectified, "it turns from the subjectivity of living presence into an object of reflective presence" (p. 34). I invited children into experiences where they could get lost in the pre-reflective moments with(in) nature that led them into reflective understandings. With(in) is a term used in this study to describe the complexities of studying nature in two ways: 1) a space outside the classroom that we go out into and work in and 2) a thing that we can feel connected to and learn with or amongst with a recognition that as humans we are nature too.

## Writing a phenomenological text

van Manen (2014) asserts that, "Phenomenological writing not only finds its starting point in wonder, it must also induce wonder … .to 'lead' the way to human understanding, it must lead the reader to wonder" (p. 360). What about nature drives us to wonder? What kind of experience draws us to silence? What kind of experience leaves us speechless? As a teacher, part of me wanted to capture everything, a running record, "my proof"—to state mandates, administration, and parents—that children were learning. However, as a researcher, I remembered to consider, what about this experience was important? How much do we capture until we know enough about the phenomena? When will we know?

Finlay (2008) suggests that an alternative to separating one's pre-assumptions is to practice with a phenomenological attitude. This means working with a process of teeter-totting back and forth from personal assumptions and returning to look at participants' experiences in a fresh way. Everything we do can have a reflective phenomenological interpretation attached to it; critical decisions were made about what experiences to "bring to life" while maintaining a phenomenological attitude. Videos, photographs, and reflection invitations, as adaptations of The Mosaic Approach, were used as potential ways to understand children's reflections. These methods do not expose the explicit meanings of one specific person, but rather suggest that these meanings are an example of a possible human experience—in our case constructed as a group. Furthermore, the intent of writing was to produce a collection of anecdotes, "that resonate and make intelligible the kinds of meanings that we seem to recognize in life as we live it" (van Manen, 2014, p. 221).

## *Punctum as good storytelling*

Anecdotes are memorable when they have a "punchy last line" or punctum (van Manen, 2014, pp. 250–252). The use of anecdotes in early childhood education is not a new method of observation (Carr, 2000; Clark, 2004, 2007; Clark & Moss, 2005, 2008). Partnered with methods of phenomenology, "anecdotes recreate experiences, but now already in a transcended (focused, condensed, intensified, oriented, and narrative) form" (van Manen, 2014, p. 250). When inviting children to reflect on their experiences, their words partnered with my narrative interpretation became anecdotes of our experience. In this study, photographs of children's pre-reflective experiences and reflective drawings are paired and placed strategically in the anecdotes to leave a lasting impression on the reader.

## Nature can be dead and alive

### *Research context: the Log Playground Tree*

The children named the forest *The Log Playground* because of one special tree. This 80-foot fallen tree with busted branches and rotted holes along its trunk afforded children spaces to climb, swing, crawl, slide, and jump like the playground at school. It was snapped off from its stump, leaving a jagged tree stump for climbing and hiding inside. One branch curved up from the trunk in a hook shape so that children could shimmy up and hang. The wider part of the tree was higher from the ground but dipped perfectly for children to slide down. The length of the tree challenged the children to balance and make it from end to end without touching the ground beneath. Children often sat on the trunk with legs at the sides creating imaginative scenarios of riding horses and dinosaurs or boarding a train and firetruck. In the following episode, Luis and Fitz contemplate an imaginative scenario around the growing forest floor, Pachysandra (see Figure 9.1).

### *Pat Cassandra and the Ninja Turtles*

Luis shouted, "Here's the Pat Cassandra!"
Fitz asked, "Who's Cassandra?"
"It's here. It's everywhere."
A smile spread across his face, "We planted Pat Cassandra around our new house at home. It has big leaves that grow and grow."
"I didn't know it was called that." I smiled back.
"My mom said, 'It'll take a while.'"
I confirmed, "It definitely wasn't this tall the last time we were here."
We walked to the edge of the Pachysandra and Fitz said "Midas, I'm Rafael!"
"I'm Mikey!" replied Midas.
Lee said, "Yeah! We're playing Ninja Turtles," as they walked land unvisited yet this year.

Suddenly Lee said, "Guys! Be quiet! I see cheek-munks."
"They're actually chipmunks," laughed Fitz.
"Cheek-munks! Ha ha!" giggled Lee.
Luis added, "Yeah, like your face!"
Midas sputtered, "No like your butt!"

**FIGURE 9.1**   Distant view Pachysandra terrain
*Source*: Adonia Porto

They all laughed and tiptoed towards the chipmunk.

Lee said with a sputter, "They're so … SO hard to see. They're SO good at hiding."

Luis asked me, "Why do they hide in the ground?"

"Maybe they go in the hole sewer like Mikey and Rafael," suggested Midas.

Luis said, "Yeah and eat pizza."

"No. They're so tiny and have to hide from predators," reminded Lee.

Zeke joined us, "Hey guys! Can I play Ninja Turtles? I'm Donatello."

"Yeah!" agreed Fitz.

"Wow guys," announced Luis, "Pat Cassandra is all over actually."

Lee asked, "Does Pat Cassandra know Ninja Turtles?"

Luis smiled, "Yep, she does!"

Sitting in the Pachysandra, Midas said, "Guys, let's make a plan to kill Shredder."

"I can fight with my hands. I don't have swinging things," Fitz added.

Midas said, "I can fight with my hands too!" and raised up his fists.

Fitz backed up and asked, "Yeah, but we're just pretending, right?"

"Let's kill Shredder in the Pat Cassandra!" suggested Luis.

Midas asked in the same breath, "Who's Pat? I don't see Pat. Is Pat your pretend friend? Do they like Shredder? Oh, is Pat a bad guy?"

"Yeah, he's a bad guy! Let's fight him!" (see Figure 9.2)

The boys immediately used fists to punch the air. Zeke shouted fighting sounds, "Hi-yah! Ugh! Cha!" and the boys copied. Then Luis announced, "The Pat Cassandra keeps growing, it's poisoning the ground. See how far it goes?" and pointed over his shoulder. Fitz,

FIGURE 9.2  Pat Cassandra and the Ninja Turtles

Midas, and Zeke continued to air fight and added some kicks to include the ground. Luis reiterated, "Everything that's green is poison and the Ninja Turtles have to be brave and get through it, then fight Shredder." Fitz nodded and shouted, "Turtles together! Turtles, let's fight!" At the edge of the green Fitz, Lee, Midas, and Zeke all carried long branches as walking sticks. They clapped their branches together and Lee confirmed, "Defeated!"

Luis' drawing as reflective response: I'm walking through the Pat Cassandra. We're the Ninja Turtles trying to fight off the poison.

Luis opened with, "Here's the Pat Cassandra!" Since he told me about the planting experience he had with his mom, I knew he was talking about Pachysandra. However, children's conversations and the Ninja Turtle context developed so quickly, I was unable to clarify. I learned later that any clarification would have been a direct interruption for what was to come. He introduced what sounded like the name of a person and the children grappled with trying to figure out who or what he was talking about while also wanting to engage in a familiar play theme, Ninja Turtles. The Teenage Mutant Ninja Turtles, a familiar knowledge of American pop culture, was something they could easily agree to enact, blending both real and imaginary realities. Immediately roles were assigned and the fighting ensued. At the same time, Lee discovered a chipmunk, calling it a "cheek-munk," a fun play on words. Hearing "cheek" led the boys to think of the cheek of their faces and bottoms causing joy and laughter. As the play unfolded, Luis continued to bring up Pat Cassandra which led the boys to inquire about it but not fully understand he was referring to the plant until later when it became the green poison covering the ground. These children had only been experimenting with language for four to five years; it made sense that they were playing with words and learning the nuances of sounds while also knowing that saying "butt" when you're four and five is funny.

Pachysandra sounds unusual when pronounced and replacing it with a person's name made it more tangible for children. Pachysandra, an evergreen perennial planted as ground cover, covered the back yard of a home on the edge of The Log Playground forest. It spreads as an invasive species, which is ironic as the boys' play considered it poison. The shocking green color of the plant and Luis' new knowledge of Pachysandra enticed the boys to move beyond The Log Playground Tree and explore a new space. Even as Ninja Turtles, Luis' interest continued to be recognized by the group as they asked questions and entertained his need to talk about it. Although playing Ninja Turtles was a prominent play theme, nature prevailed in the children's experiences. Pachysandra became a play space simply because it was eye capturing green and different than the surrounding dead leaves (see Figures 9.1 and 9.2). Pat was also a *she* that knew Ninja Turtles before becoming a *he* that was a bad guy and ultimately poison spreading over the ground. With each revisit of attempting to understand Pachysandra, the complexity of their understandings grew.

## The potato bug

Past the Pachysandra, Lee led them to a pile of fallen trees and he swiped his stick on a
  log. "Hey, I found something!" He held it in his hand.
Midas asked him, "Is it a roly poly?"
"Nope. It's longer. A millipede, I think." replied Lee.
"Millipedes curl when they're scared and stay like that awhile. Once they like you, they
  uncurl," he continued.
I asked, "Can anyone else find something that's alive?"
Luis remarked, "Pat Cassandra is alive, see?" and pointed all around.
"Yeah, you are right. It seems to be in full bloom right now."
Zeke said, "I found a leaf."
"What about that leaf is alive?" I asked.
He looked close and said, "Well if you look really close you can see the lines where it
  used to breathe. It's dead now."
"Nature can be dead and alive," announced Luis.
I asked, "How do you know?"
Fitz said, "Well if something used to be alive then it's nature even if its dead."
I nodded, "What else can you tell me about that?"
They both sputtered and I said, "Wait a second. Let Fitz finish."
He said as he grabbed his chest, "When you breathe in, you're alive. When you stop
  breathing, you're dead. Nature breathes and then it's dead."
Luis nodded, "Yeah. Leaves breathe in the summer but die in winter. New leaves
  come."
"Yep. I knew it," agreed Lee.
I asked, "Lee, do you want to add anything?"
"Well, I used to think worms never died but then I saw them on the black path and
  Midas killed one. More worms were squished too."
Midas said, "Maybe I killed it on accident, maybe I didn't."
Zeke asked, "Did more worms come?"
"When it rains, the worms kind of drown and come out and die or live. When it doesn't
  rain, they're happy in the mud," Lee said as he uncurled a millipede in his hand.
I asked, "Do you know why those curl up like that?"

"Yeah, 'cuz it's scared."

Zeke held a roly poly bug, "Mine is curled."

I asked, "Is yours scared, too?"

"Yep, it's a roly poly."

Lee said, "Well my grandparents and my mom and dad call that a potato bug."

Zeke popped the bug into his mouth and within three seconds spit it out!

He shouted, "Well you can't eat it like a potato!"

"EWW! Don't eat it. They just call it that because it's an ov-val like a potato," remarked Lee.

Zeke firmly stated, "It's not a potato bug, it's a roly poly."

At the same time, Rosie called from a distance, "We found ear worms over here! Wanna come see?" (Figure 9.3)

Midas, Lee, Fitz, Zeke, and Luis were contemplating the question, what is nature? First, Lee described how the millipede was fearful of him but would uncurl when it was no longer afraid. The defense mechanism of a millipede is curling to protect itself from predators. When the millipede did not uncurl on its own, Lee helped it (see Figure 9.3, 9.4, 9.5). Lee forced the millipede to "like" him.

Carefully asking about other things "alive" led the boys to consider that nature can be dead and alive. Zeke talked about how the lines are how leaves breathe. Fitz showed us breathing, letting us understand that if something used to be alive then it is nature. Luis added the transformation of seasons and the coming of new leaves. I assumed that Lee also realized that all nature dies as he described how he used to think worms never died which encouraged us to see again Midas' desire to fit in with the group but his tendencies to harm nature. Anyone who has ever killed an insect, perhaps in their home, understands Midas' desire. This anecdote reminds us that to phenomenologically understand we must include moments about nature and moments about something else as the Pachysandra led Ninja Turtles to fight it as poison but then understand that nature can be dead and alive.

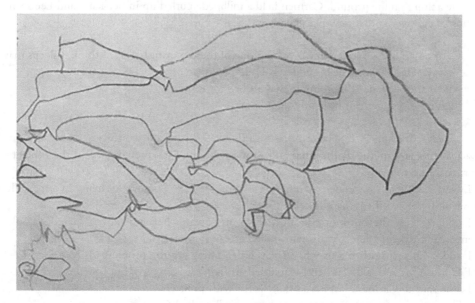

**FIGURE 9.3** Lee sharing his knowledge of curled millipedes

*Source*: Adonia Porto

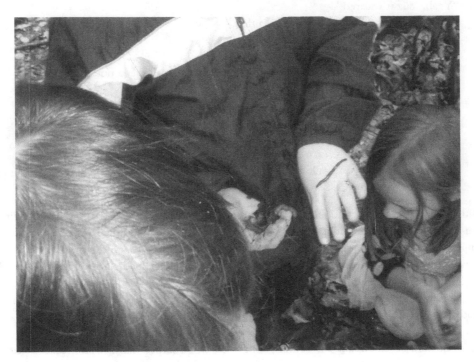

**FIGURE 9.4** Helping his millipede uncurl
*Source*: Adonia Porto

## Ear worms

The group headed back towards The Log Playground Tree where Carmen, Nikki, and Milly were searching on the ground. Carmen held a millipede curled up in her hand and Lee shared, "Those were over in that new place we went to. It's scared."

> Carmen argued, "No it's not. I've seen one of these worms before. I think it sleeps that way. That's how it sleeps underground."
> Midas announced, "I'm going to make a bed for mine."
> "Let me try to get this little guy some food."

Carmen's drawing as reflective response (Figure 9.6): "these worms are eating the mud. Two worms are smiling at the mud and eating it because it's their food and the other two worms are best friends."

Lillian's drawing as reflective response (Figure 9.7): "There's so many worms. They move and wiggle off my hand like all the time."

I noticed Tess, the outdoor educator, wave me towards her and she whispered, "Look at them taking care of these worms with such care and empathy."

I said, "This reminds me of when I was little. These are the moments I lived for."

Fitz heard me and said, "When I was little I wouldn't hold worms but now I do." I smiled.

Rosie said to Carmen, "Where's that worm with pointys?"

"That wasn't a worm. Worms don't have ears," replied Carmen.

Milly added, "Yep. It was a caterpillar."

**FIGURE 9.5** The tasted potato bug

*Source*: Adonia Porto

Rosie moved a pile of leaves and discovered a large earthworm and screeched, "Worms, everybody!"

Lillian picked up a small pile of them and let them move in her hand while Rosie touched several and with a squeaky voice mumbled, "A thousand."

Milly pleaded, "Can I have a worm because I don't have a worm?"

Tired of waiting she moved a leaf and found her own. Then Rosie snatched up something small and hid it in her hand right away.

I said, "What did you find, Rosie?"

"One of those ear worms again and it's MINE!"

I asked to see it and she said, "But don't give it to someone else. It's curled 'cuz it's scared but it does have little ears."

Nikki remarked, "It hears you say that. You should tell him a bedtime story."

**FIGURE 9.6** Carmen's drawing as reflective response: "these worms are eating the mud."
*Source*: Adonia Porto

Land millipedes are a species of millipedes that have antennas. The children referred to them as ears and named them ear worms. Then Nikki dangled a worm in Zeke's face and said, "Don't let my worms cry."

He shrugged and said, "We're gonna cook them up some dirt. Let's go."

The group moved the worms to rest on The Log Playground Tree. Carmen, Milly, and Nikki made worm beds including pillows and blankets. Carmen also made one leaf into a pot of dirt for worm food. Midas' bed became a pile of five worms and he shared, "Mine are getting cozy. They like their bed." Once the worms were all tucked in, we headed back to school.

## In-seeing: nature can be dead and alive

In this phenomenological experience, the children became nature as the roly polies and millipedes (ear worms) were personified to need things that children need (like cosy beds

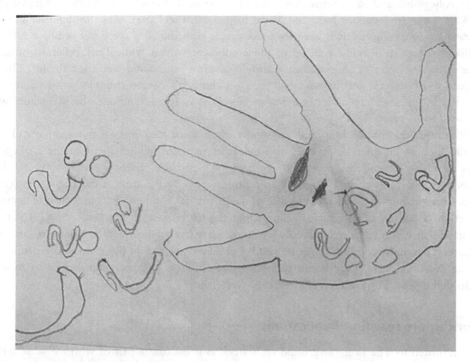

**FIGURE 9.7** Lillian's drawing as reflective response: "there's so many worms. They move and wiggle off my hand like all the time."
*Source*: Adonia Porto

**FIGURE 9.8** Worm's cosy bed
*Source*: Adonia Porto

and blanketes). Children recognized that nature is like them and nature became them. The roly polies and millipedes were curling with scared feelings and sleeping in their beds. They needed bedtime stories and food and heard what the children were saying to them. They were served pillows and blankets for beds and pots of dirt for food. The children were caring for them as we (teachers) cared for the children, as we cared for each other.

The roly polies and ear worms may have cried without proper care, Nikki reminded us. Zeke's "little guy" needed food and Midas helped them feel cozy. Zeke thought he'd try a potato bug, meeting his own need for food, before realizing they were not ideal for eating. Lillian's reflective drawing referred to the number of worms while Luis' reflected on the size of the Pachysandra. Additionally, Carmen's worms smiled and were best friends. These reflections, both said and drawn, shed light on children's constructions of what nature is, can be, and how the dynamics of living species allow us to become one and like nature we all need to be cared for.

Pat Cassandra, potato bug, and ear worms were three anecdotes that were particularly challenging to experience, capture, and reflect upon with children. Part of the challenge was that the experiences happened fast and back-to-back. Each anecdote was equally important and so I worked diligently to capture them. At the time, I did not know what they were telling me but I could tell it was a pivotal moment in our study of nature together. I had an indescribable feeling about the experiences I was "in-seeing." Later, I realized that it was because children were recognizing their own needs as being the same as nature. They needed care like nature needs our care. As children began to know nature as being dead and alive, they began to realize that children had more similarities to nature than differences. The children realized they were nature.

## Conclusion: teaching implications

We returned to The Log Playground each week as it became a part of who we were with (in) nature, and our relationship with the place allowed us to (re)know nature with confidence in our routine. Revisiting the same space over time allows children to (re)know, discovering something different than the visit prior (Tovey, 2007). Uninterrupted time in nature provided children with freedom and agency to learn on their own terms and make decisions about their connections with(in) nature (Moss & Petrie, 2002; Waller, 2008). The children knew that I was there to record their moments in nature with the intention to understand how they learned about nature and how they interacted with earth. They knew that if they needed my support in any way they could ask for a conversation, a tissue, a bandage, or a break away from the group. Creating this learning context together allowed children to move freely with a sense of wonder and grasp nature for what it was during our time together without an adult research agenda. However, as a teacher with obligations at our school, I did have to make sure we made it back to school in time for lunch.

In order to fully understand children's constructions of nature's phenomena, I diligently considered my own phenomenological attitude whenever children invited me into their conversation. *Phenomenological listening* is a phrase I use to describe practicing listening methods while maintaining a phenomenological attitude. I was constantly checking in with own my interpretations and reflections before responding or posing any questions with children or including them in the written anecdotes. It was important to consciously monitor my thoughts while recording and being in the moment with children because I truly wanted to understand how our time in the space let children know nature differently. While the methodology of this research was challenging at times, when needing or wanting to intervene for various reasons, trusting the children's ways of being and becoming with (in) nature proved most significant to the group's learning (Davies, 2014).

Listening to children in The Log Playground opens several curriculum realities for teachers wanting to approach curriculum in the outdoors. In addition to providing uninterrupted time in nature with a listening context, teachers should consider the following.

## Children's grappling with literal understandings

In these three anecdotes children grappled with language understandings: cheek-munks, Pat Cassandra vs Pachysandra, and Potato Bug. The play on words forced them to grapple with word meanings and literal meanings of what the group was trying to teach one another. They also grappled with significant life events such as Good vs. Evil and Life and Death. Once children began to realize that nature was "breathing" and that "nature needs bedtime stories," children noticed how their actions could impact earth in negative and positive ways. Good vs. Evil comes out in many pop culture avenues which gave the children a way to deal with a large field of Pachysandra that felt overtaking of one of their preferred natural environments. Ironically, as a group we decided not to intentionally harm nature as we learned it needed our care, so in this instance, the boys were tip-toeing over the living invasive species while symbolically attacking it as poison. Life and Death emerged several times in the anecdotes as children came to understand that if something used to be alive, it breathed, as we breathe. Midas learned the power of his actions when Lee described them which may have left an impression on his view of killing insects as a way of being with(in) nature. Supporting children's grappling with these life events provides them with the skill set to conceptualize how others view the world, but more importantly, how they want to view it for themselves.

## Visually capturing children's experiences

Adapting the magic carpet method of The Mosaic Approach and using video recordings of pre-reflective moments encouraged us to relive our in-the-moment experiences for a second time and prompt a stronger reflection. Observing children watching their own pre-reflective experiences was the best example of "in-seeing" I could fathom (van Manen, 2014). Children expressed shock, thrill, laughter, and sometimes sadness about how they interacted with(in) nature in the videos. When Fitz saw himself holding a giant earthworm, he still felt shocked by viewing the size of the worm and wetness he described as slimy when touched in the video. This aligned with his direct sentiment earlier that "he didn't hold worms when he was little." When Zeke viewed himself tasting a potato bug, he actually coughed in disgust. He described his pre-reflective experience as, "not even thinking about it." Furthermore, when Nikki watched herself on video say, "Don't let my worm cry" to Zeke, she looked as though her empathy was so strong she might cry in that moment. When I asked her if she was okay, she said, "I'm a little sad." Children's emotions to nature were strong and vivid evoking many complex responses.

## Encouraging children to draw as reflection

Mapmaking, or children's two-dimensional drawings about their experiences, was another valuable method of The Mosaic Approach that encouraged children's reflections of their experiences beyond me simply asking, "What was it like to be in the space or with that part of nature?" Encouraging children's drawings in our study together prompted children's thinking in a different way and prompted a stronger reflection than my questioning would promote. Carmen's drawing described how worms were happy to see their food and have best friends, which allowed us to understand that worms need food and best friends just as humans do. Luis' drawing suggested his way of understanding how much Pachysandra was at The Log Playground that day (his lines filled the piece of paper) and the impact of the

Ninja Turtle play theme and fighting poison. Children's drawings were a strong narrative tool that encouraged children to draw what they found significant about their experience (Clark & Moss, 2008; MacDonald, 2009, see Figures 9.3, 9.5).

These curriculum realities offer potential ways of being with children and knowing children and nature differently. Listening to children with these intentions will provide children with agency to make decisions about being and becoming in learning contexts, how knowledge is constructed, individually or in a group, and how they connect with(in) nature. More importantly, if children connect with(in) nature on their own terms, their relationship with nature, nature both dead and alive, will last into adulthood and has the potential to encourage future sustainable efforts to preserve nature and earth as we know it.

## References

Carr, M. (2000). Seeking children's perspectives about their learning. In A. Smith, N.J. Taylor, & M. Gollop (Eds.), *Children's voices: Research, policy and practice*, 37–55. Auckland: Pearson Education.

Chawla, L. (1998). Significant life experiences revisited: A review of research on sources of environmental sensitivity. *The Journal of Environmental Education*, *29*(3), 11–21.

Chawla, L. (2007). Childhood experiences associated with care for the natural world: A theoretical framework for empirical results. *Children, Youth and Environments*, *17*(4), 144–170.

Clark, A. (2004). The Mosaic approach and research with young children. In V. Lewis, M. Kellett, C. Robinson, S. Fraser, & S. Dingm (Eds.), *The reality of research with children and young people*, 142–156. London: Sage.

Clark, A. (2007). A hundred ways of listening: Gathering children's perspectives of their early childhood. *Young Children*, *62*(3), 76–81.

Clark, A., & Moss, P. (2005). *Spaces to play: More listening to young children using the Mosaic approach*. London: National Children's Bureau.

Clark, A., & Moss, P. (2008). *Listening to young children: The mosaic approach*. London: National Children's Bureau.

Copeland, K. A., Kendeigh, C. A., Saelens, B. E., Kalkwarf, H. J., & Sherman, S. N. (2012). Physical activity in child-care centers: Do teachers hold the key to the playground? *Health Education Research*, *27*, 81–100.

Corsaro, W. (1997). *The sociology of childhood*. Thousand Oaks, CA: Pine Forge Press.

Davies, B. (2014). *Listening to children: Being and becoming*. London and New York: Routledge.

Demeritt, D. (2001). Being constructive about nature. In N. Castree & B. Braun (Eds.) *Social nature: Theory, practice and politics* 22–40. Malden, MA: Blackwell.

Dyment, J., & O'Connell, T. (2013). The impact of playground design on play choices and behaviors of preschool children. *Children's Geographies*, *11*(3), 263–280.

Finlay, L. (2008). A dance between the reduction and reflexivity: Explicating the "phenomenological psychological attitude". *Journal of Phenomenological Psychology*, *39*, 1–32.

Fjortoft, I. (2001). The natural environment as a playground for children: The impact of outdoor play activities in pre-primary school children. *Early Childhood Education Journal*, *29*(2), 111–117.

Fjortoft, I. (2004). Landscape as playscape: The effects of natural environments on children's play and motor development. *Children, Youth and Environments*, *14*(2), 23–44.

Fjortoft, I. & Sageie, J. (2000). The natural environment as a playground for children: Landscape description and analyses of a natural playscape. *Landscape and Urban Planning*, *48*, 83–97.

Kahn, P., & Kellert, S. (2002). *Children and nature: Psychological, sociocultural and evolutionary investigations*. Cambridge, MA: MIT Press.

Kellert, S.R. (2002). *Experiencing nature: Affective, cognitive, and evaluative development in children*. In P. Kahn & S. Kellert (Eds.) *Children and nature: Psychological, sociocultural and evolutionary investigations*, 117–152. Cambridge, MA: MIT Press.

Kernan, M., & Devine, D. (2010). Being confined within? Constructions of the good childhood and outdoor play in early childhood education and care settings in Ireland. *Children & Society*, *24*, 371–385.

Little, H., & Eager, D. (2012). Risk, challenge and safety: Implications for play quality and playground design. *European Early Childhood Education Research Journal, 18*(4), 497–513.

Little, H., & Wyver, S. (2008). Outdoor play: Does avoiding the risks reduce the benefits?. *Australian Journal of Early Childhood, 33*(2), 33–40.

Little, H., Wyver, S., & Gibson, F. (2011). The influence of play context and adult attitudes on young children's physical risk-taking during outdoor play. *European Early Childhood Education Research Journal, 19*(1), 113–131.

MacDonald, A. (2009). Drawing stories: The power of children's drawings to communicate the lived experiences of starting school. *Australian Journal of Early Childhood, 34*(3), 40–49.

MacNaughton, G. (2003). *Shaping early childhood: Learners, curriculum and contexts.* Maidenhead, UK: Open University Press.

Moss, P., & Petrie, P. (2002). *From children's services to children's spaces.* London: Routledge Falmer.

Murray, R., & O'Brien, E. (2005). *Such enthusiasm-a joy to see: An evaluation of Forest School in England.* Report to the Forestry Commission by the New Economics Foundation and Forest Research.

O'Brien, E., & Murray, R. (2006). *A marvelous opportunity for children to learn: A participatory evaluation of Forest School in England and Wales.* England: Forestry Commission.

Rinaldi, C. (2006). *In dialogue with Reggio Emilia: Listening, researching, and learning.* New York: Routledge.

Sobel, D. (2008). *Childhood and nature: Design principles for educators.* Portland, ME: Stenhouse.

Staempfli, M. B. (2009). Reintroducing adventure into children's outdoor play environments. *Environment and Behavior, 41*(2), 268–280.

Stan, I., & Humberstone, B. (2011). An ethnography of the outdoor classroom-how teachers manage risk in the outdoors. *Ethnography and Education, 6*(2), 213–228.

Tovey, H. (2007). *Playing outdoors: Spaces and places, risk and challenge.* Maidenhead: McGraw Hill.

van Manen, M. (2014). *Phenomenology of practice: Meaning-giving methods in phenomenological research and writing.* Walnut Creek, CA: Left Coast Press.

van Manen, M. (1984). Practicing phenomenological writing. *Phenomenology + Pedagogy, 2*(1), 36–69.

Waller, T. (2008). "The trampoline tree and the swamp monster with 18 heads": Outdoor play in the foundation stage and foundation phase. *International Journal of Primary, Elementary and Early Years Education, 35*(4), 3–13.

# 10

# IMAGINE SUSTAINABLE FUTURES

## Experimental encounters between young children and vibrant recycled matter

*Nina Odegard*

> In trying to become 'objective,' Western culture made 'objects' of things and people when it distanced itself from them, thereby losing 'touch' with them.
>
> Gloria E. Anzaldúa (1987, p. 59)

## Introduction

Academics in various disciplines have taken up Crutzen's (2006) term *Anthropocene* (the era since the Industrial Revolution) to address the escalation of the human-made planetary problems and express humans' entanglements with the fate of the Earth (Colebrook, 2012; Somerville, 2013, 2015). Acknowledging these humans' entanglements opens the possibilities for emergent responses to connect nature and culture, economy and ecology, as well as natural and human sciences, in a more sustainable way (Somerville, 2013). How can we make these sustainable connections in early childhood education and care (ECEC)?

I try to move closer to sustainable connections by thinking with the feminist posthumanist philosophies of Barad (2008, 2012), Duhn (2012, 2015, 2016; Duhn & Quinones, 2018), Haraway (2016), and Lenz Taguchi (2017). Language has gained excessive power in research; Barad (2008) claims, "[T]he only thing that doesn't seem to matter is matter" (2008, p. 103). Matter is understood as a "material substance that occupies space, has mass, and is composed predominantly of atoms consisting of protons, neutrons, and electrons, that constitutes the observable universe, and that is interconvertible with energy" (Matter, 2018). Thus, materiality and meaning do not comprise separate elements but are entangled and mutually dependent. Drawing on a new materialist theoretical framework that emphasizes the agential qualities of matter, sustainability, and pedagogy in ECEC, I argue that recycled materials possess other potential than simply being commodities that humans use and throw away (Kind, 2014, p. 868).

The renewed focus on materiality "is a consequence of the disruption of the human-centrism and the questioning of dualisms such as matter/meaning, mind/body, subject/object, and nature/culture" (Ceder, 2016, p. 61). New materialism questions the dualism of the active human subject and the passive material object. It views matter as vibrant

(Bennett, 2010) or agential (Barad, 2008) and "rules out the possibility for humans to think [of] themselves as removed from a multitude of material agencies" (Alaimo, 2018, p. 52).

By including the nonhuman in questions about who matters and what counts, posthuman research practices may offer new ethics of engagement in education (Taylor, 2016, p. 5). Hence, I study the aesthetic encounters among humans, recycled materials, and materialities using Manning's (2013) *objectiles* and Bennett's (2010) *vibrant matter*. By putting these two intrarelated concepts to work, on one hand, Manning's objectiles allow me to theorize about the (active) multiplicity of objects and their field of relations, which are never disconnected from the event. "An object is as much how it does as what it does" (Manning, 2013, p. 32). On the other hand, Bennett's vibrant matter allows me to highlight the contours of how more-than-humans assemble.

I draw on an ongoing research project that studies young children's aesthetic explorations with recycled materials in a creative reuse center, called a *Remida* (Vecchi, 2010, 2012). Aesthetic explorations have formed a central concept in this research and have developed as continuously multisensory, explorative, and aesthetic (as in inviting, becoming, and thinking) encounters, crossing "borders and timelines where play and learning are intertwined" (Odegard & Rossholt, 2016, p. 55).

Research questions often decide on or define the methods, as well as address the collaborators/agents with whom the researcher cooperates. This study has taken the form of a visual and sensory ethnographic research project (Pink, 2007, 2015). The qualitative data include images and video recordings, field notes, and transcripts of talks with the adults after each session (called after-talks). I have also employed different ways of writing, such as the visual writings used in this chapter, which "write themselves across subject matters, discourses, genres, images, and other objects and practices" (Ulmer & Koro-Ljungberg, 2015, p. 141).

By unwrapping vibrant matter and objectiles as theoretical concepts and engagement in young children and recycled materials as common worlds, I inquire how this undertaking can contribute to ideas, thoughts, and imaginings for sustainable futures.

## The Remida and working with recycled materials

The international Remidas (around 18 and growing) are all inspired by the first one, founded in 1996 as a project of infant-toddler centers and preschools in the Municipality of Reggio Emilia; North Italy (Remida, 2018). Each Remida is closely connected to democratic ideas, values, and principles, collectively known as the Reggio Emilia approach. A common ground is that the centers are usually based on cultural and pedagogical projects that encourage optimistic and proactive ways of approaching environmentalism (Remida, 2018; Vecchi, 2010, 2012).

In addition to the formal Remidas, many creative recycling and reuse centers exist. The Nordic countries have approximately 30 established creative reuse centers besides several Remidas, with the growing trend of establishing educational and environmental spaces where we can learn about recycled materials. These centers (Remidas and creative reuse centers) have a large variety of materials and "are partnered with manufacturers to reuse and repurpose materials discarded from the end-product manufacturing process, keeping thousands of pounds of material out of landfills" (Parnell, Downs, & Cullen, 2017, p. 236). The centers' employees collect, store, and exhibit unique materials, so children, pupils, educators, artists, families, and community members "can see what possibilities repurposing material create[s]" (Parnell et al., 2017, p. 236). Each center offers different materials, relying on the variety of industries and businesses in the local area.

In recent articles, I have explored whether recycled materials, with their vast variations of textures, shapes, colors, and sizes, evoke curiosity and creativity (Odegard, 2012; Odegard & Rossholt, 2016). Through different arrangements and cooperation with teachers and artists in workshops, lectures, and exhibitions, a creative recycling center proactively attempts to initiate changes and a new optimistic view by giving value to recycled materials (Odegard, 2015; Remida, 2018).

In an interview with Carlsen (2013, 2015), Reggio Emilia's *atelierista*[1] Vea Vecchi underlines the importance of knowing that Remida materials are sourced from industrial wastes. According to her, materials retrieve significance and meaning from their specific contexts, and the materials' origins change how they are viewed (Carlsen, 2013, p. 23). Supposedly, waste materials, such as recycled items, connect to the global changes that challenge our ways of living and being. Knowing our waste materials, determining their status and existence through sorting, and exhibiting and offering them to visitors, constitute a central part of my work as a project leader of a creative reuse center. Exploring with these materials is also a way of acknowledging their history, value, and supposed destiny and of learning how to "stay with the trouble" on our damaged Earth (Haraway, 2016). How these materials are valued depends on knowing about them: "anything and everything can become waste, and things can simultaneously be and not be waste, depending on the perceiver" (Hird, 2012, p. 454).

In the following sections, I work with the concepts of vibrant matter (Bennett, 2010) and objectiles (Manning, 2013) in relation to recycled materials and discuss these materials' agency, vitality, and capacity to make things happen.

## Thinking with vibrant matter and objectiles in young children's aesthetic explorations

In the early years' setting and the creative reuse centers, I have experienced how the children's connectedness with the vibrancy of matter plays out (Tesar & Arndt, 2016, p. 193). Bennett (2010) encourages us to think with powers other than human, think with matter, engage with it aesthetically, and think with material powers that circulate around and within the human body, which can "aid or destroy, enrich or disable, ennoble or degrade us, in any case, call for our attentiveness" (2010, p. ix). With this theory of vibrant matter, Bennett (2010) argues that objects possess a *thingly power* (with material agencies) that is "vital, vibrant and impermanent" (Ceder, 2016, p. 110). A thingly power behaves in unpredictable ways (Bennett, 2010) and appears in assemblages where forces and things impact and shape other matters and things (Tesar & Arndt, 2016, p. 193).

In my earlier work (Odegard, 2012), I have discussed that recycled materials seem to *lose their function* in relation to children's aesthetic explorations. Manning's (2013) use of the objectile concept appears similar, yet it adds something more. An object becomes an objectile, when we think less about what the object is and more about what an object does, about its "capacity to generate event-time" (Manning, 2013, p. 92). Here, event-time is understood as what is felt in the moment and how an object is already in a field of relations (Manning, 2016). Manning builds her understanding on Deleuze's (1993, p. 19) work that uses the term *objectile* to emphasize "not only a temporal but also a qualitative conception of the object, to the extent that sounds and colors are flexible and taken in modulation." Manning's work as a choreographer seems to concentrate on objectiles as recognizable everyday objects, already known to their user. In contrast, recycled materials often have unknown and often unrecognizable properties. However, recycled materials are comparable to objectiles in terms of their provoking, inviting, and suggesting properties and in how they act in a field of relations.

In the following sections, I briefly present the research methodology. The images and the visual writings are carefully selected to illustrate recycled materials as vibrant matter and objectiles.

## Research field, methods, and methodology: some insights

The research was conducted in one of the Nordic Remidas. Thick black curtains surrounded the middle of the room and made a rectangular space called the *blackbox*. In this space, young children and pupils were invited and encouraged to experiment and explore recycled materials with various analog and digital tools, such as spotlights on the ceiling, a projection screen, a projector, an overhead projector, light tables, and a white plexiglass sheet on the floor, among other tools.

While I conducted the study, I asked if the artist could contact the early childhood centers (ECCs) that had recycled materials in their environments, as well as the children who had previously visited the Remida's blackbox. The artist, who was also the center's leader, organized the groups and managed all the sessions. The young children were mostly from the ECCs whose work was inspired by the pedagogy from Reggio Emilia, and they were organized in small groups of three to five children. The atelierista from the nearby ECC was asked to join the research on the grounds of her considerable knowledge about working with young children and this Remida's materials.

With the exception of the artist, the adults (including the teachers, the atelierista, and myself) withdrew from any action but responded if the young children invited us in or needed support. As the researcher, I was an observer through the camera lens but was sometimes invited by the children to join their explorations and play since I (similar to the other adults) was sitting and available to them in the blackbox. This approach made me feel connected and disconnected at the same time. This methodology made it possible to study the encounters among the recycled materials, the children, and the materialities, and it was also used out of respect for the children's group construction of understanding these open-ended materials. This distance also shifted my attention from following the humans to what was going on between the humans and the non-humans and between the matter and the materialities, which was central in this research. After-talks with the adults followed each session.

As mentioned, this fieldwork took the form of a visual and sensory ethnography where I photographed, video recorded, and wrote field notes while observing. This approach departs from the classical observational type and is a more critical methodology that opens up to "multiple ways of knowing and exploration of and reflection on new routes to knowledge" (Pink, 2015, p. 5). Through Pink's (2015) interpretation, visual and sensory ethnography "does not claim an objective or truthful account of reality" (2015, p. 5) but aims to be loyal to the ethnographer's embodied, sensory, and affective experiences. The qualitative data in this research consisted of transcripts from video recordings and after-talks, photos, field notes, and different ways of writing. Through three articles, I have worked with the data from this research in different ways, for example, in a rhizomatic analysis (Odegard & Rossholt, 2016) and through the concept of hypermodality (Odegard, 2019). In this chapter, I approach the data from specific events to illustrate and discuss two theories through their concepts of objectiles and vibrant matter.

## *Aesthetic explorations of recycled matter*

To illustrate the creativity of thought through the concepts of vibrant matter and objectiles, as well as collective creative processes, I used the following events documented in photos, transcribed videos, field notes, after-talks, and visual writings. These events involved a small group of young children, their teacher, the artist, and the researcher, as illustrated in the following paragraphs. Non-human objects, including a white plexiglass sheet, different industrial metal objects (as presented in the images), frames with transparent colored sheets (for stage lighting, originally called gel frames/light frames), and light tables, were also available. By viewing and transcribing videos, viewing photos, listening to and transcribing after-talks, and going through different notes, I engaged with each of these encounters. In an embodied, sensory, and affective way, I aesthetically explored the events and introduce them in this chapter as summarized narratives with qualities that may express their aliveness in the experiences with the children.

In this section, I present three events: four images, three texts, and three visual writings. These particular events picked me, pulled me in, made me halt, wonder, and question them through their materialities, colors, constructions, and beauty. Pictures, texts from my new encounters with visual elements (video and photo data), and the visual writings illustrate the events. These images and texts belong to a larger dataset from my doctoral research but are used in this chapter to illustrate and underline the connections between the concepts of vibrant matter and objectiles, along with the imaginings they may offer.

First, I introduce the narratives with images, texts, and visual writings. Second, I employ each event, its texts, and visual writings to explore their relations with vibrant matter and objectiles.

**FIGURE 10.1** In the video, five young children encounter the recycled materials they are holding
*Source*: Nina Odegard

In the video, five young children encounter the recycled materials they are holding, feeling the weight of the items in their hands, studying them with their eyes, looking at them from different perspectives, and feeling them in different ways against their skin, clothes, and hair.

Furthermore, the children seem to explore the potentials of the metal surplus materials on the floor by building various pieces of metal on top of each other at different levels. They watch one another attentively and try out different techniques to balance the different metal pieces. The children discuss among themselves and help one another. As observers, we could sense their experimentations, as well as our common concentration and attentiveness.

> Strange unknown objects in red light. Red light, shadows, hands, sounds of metal encounters … Who are they? What are they? Where do they come from? What were they parts of? What are they becoming? What can they do? What can they make humans do?
>
> *(Text from the video-data schema 1: (8.6.15), field notes (8.6.15), and new encounters with the data. Visual writing when encountering video sessions)*

**FIGURE 10.2**  The children use surplus materials (in this case, light filters), combine their different colors, and cover the whole light table

*Source*: Nina Odegard

The ways that the children study the light filters do not look coincidental. The children move them around, look through them, combine and select each light filter, and place them in an aesthetic manner. However, the pattern and the aesthetics do not seem to follow anything else other than the materials and the children's mutual vibrancies. The light filters seem deeply involved in the young children's building through their aesthetics.

New objects, light filters, transparent and in many colors. Human and nonhuman aesthetic explorations. What can colors do? Layers of color? Circles, hands, and squares. Frames. Light frames. Color frames. Framing beauty, tame it, or set it free?

*(Text from video-data schema 1 (8.06.15), field notes (8.06.15), and new encounters with the data. Visual writing when encountering video sessions)*

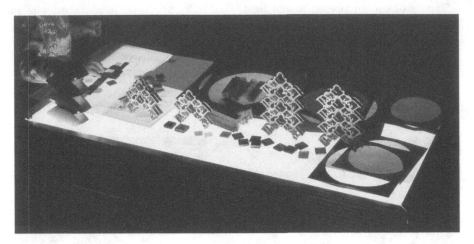

**FIGURE 10.3**   The children use surplus materials (in this case, light filters), combine their different colors, and cover the whole light table

*Source*: Nina Odegard

**FIGURE 10.4**   Two children move light filters to the other light table and start to build

*Source*: Nina Odegard

## Connecting different experiences and building a city

Two children move light filters to the other light table and start to build. First, it is one dimensional and flat, and then, without a single word spoken, they develop their construction with three-dimensional figures. They use the same metal pieces that they constructed with earlier, combining aesthetic explorations and constructions on the floor with the one on the light table. Their concentration, silence, attentive attitude toward each other, and beauty are making the teacher, the artist, and the researcher fascinated, quiet, and overwhelmed. They are aesthetically exploring the materials by using different senses in their encounters with the different objects and matter, such as light, color, and shadow.

First, they are on the floor building high constructions, and then exploring the materials' potentials. They set the light filters on the light table, finally combining their aesthetic experiences and the materials' potentialities in the combination of the two. Sitting quite close, I listen to their talk of building a city. Their story about what they are building seems insignificant; it almost seems to narrate itself while they are building.

(Text from video-data schema 1(8.06.15), field notes (8.06.15), and new encounters with the data)

> The transparency of the colors works differently when it encounters the metal objects. Changeable objects nourish new ideas. Colors in combination. Contrasts. Capacity? What can these events do? Materials in complex constructions. Alter alternatives? How do these materials affect us? And how do humans affect them? Imaginings.
>
> *(Visual writing in the encounter with the video session)*

These events seem to have a linear orientation in the way that they are presented in this chapter. However, to me, these events are entangled and shaped by the encounters and our common narratives.

## Vibrant recycled matter and objectiles

A relatively traditional way of reading this material is to develop codes or categories in order to "both reveal and identify prominent themes" (Lafton, 2015, p. 145). However, in this chapter, my aim is to open up to the complexities and the possibilities of young children's encounters with recycled materials through the lens of vibrant matter and objectiles, not to identify themes. In empirical research, the challenge lies in formulating (and maybe reformulating) the chosen research problem together with the different agents (human and non-human) and texts (Lenz Taguchi, 2017, p. 188). By re-reading the materials (data) and including contributions from human and non-human participants, I engaged with the concepts of vibrant matter and objectiles to help me unpack the entanglements in these narratives. In this particular session, children, lights, recycled materials, and materialities seemed to contribute to the construction of research entanglements and common knowledge.

In the first narrative, five children aesthetically explored metal objects (Figure 10.1). The young children's hands and bodies constantly shifted and moved with the materials, drawn on their own and by the materials' capacities and potentials. Some of the objects were shining, and different shadows were cast by the objects in their encounters with the light. The light-matter seemed to contribute also, with its shades and color, as well as how the materials responded to the light (shiny/matte/opaque). The aesthetics were made visible through the red light that covered the whole area and colored some parts of the metal objects. As I encountered this phenomenon, I wondered how objects and young children are affectively related to each other's vibrancy.

Through the concept of agency, recycled materials' contributions to the young children's aesthetic explorations in the blackbox materialized. The sceneries invited the children to "examine nonhumans more closely, to listen and respond more carefully to their outbreaks, objections, testimonies, and propositions" (Bennett, 2010, p. 108). What could happen in the field of ECEC if we would explore the idea that the materials invited the humans in and shaped them and their building? What could occur if we would accept the phenomena of the materials' agency and the proposals for humans, where recycled materials could be considered layered, multiple, and affective forces that invited us to think with them by drawing our attention? In the narratives, the vibrant recycled matter contributed with their own "language" (literacy), and the entanglements of children-materials-matter explored one another's potentialities. Through different materialities and surfaces, building properties, shapes, and how they feel against the hands (e.g., cold or warm, shaped or unshaped), the materials' agencies, possibilities, and potentials as nonhuman companions emerged.

Similar to the narratives written above, in several after-talks (transcripts), the adults expressed their awareness of how the recycled materials triggered the children's (and their teacher's) curiosities, constituting the materials' vibrant matter. They told narratives where the children seemed to be pulled into the light, the colors, the shadows, and the materials. In a particular after-talk, the teacher and the atelierista discussed the vibrancy of the white plexiglass sheet on the floor. They talked about how the white sheet appeared to gather the children and how this plexiglass functioned as a rallying point in itself. The atelierista also told the story where the plexiglass' vibrant matter seemed to invite the children to playfully tease the plexiglass in their first encounter—first with one finger, one knee, and one foot, back and forth—before they finally placed themselves in the middle of the white plexiglass sheet.

In the Remida, young children had access to different analog and digital tools, such as flashlights, light tables, an overhead projector, a projector, a digital microscope, and light equipment. Through these different technologies, Bennett (2010, p. 108) suggests that humans may pay more "attention to different matters' vibrancy" (p. 108). Both digital and analog tools opened up possibilities and imaginings, and they also appeared to strengthen the materials' vibrancy. "Thinking and being, digitization and humanness are mutually productive and intertwined. Furthermore, we are multi-sensory in our access of knowing" (Holloway-Attaway, 2018, p. 94) The human and the non-human seemed to be mutually remade in these encounters due to the reciprocity of their vibrancy.

In the second narrative, two young children covered the whole table repeatedly with the colored light filters, and then they layered them to explore new colors (Figure 10.2). These decisions were made "in real time and without a predetermined outcome" (Bennett, 2010, p. 97). The Remida's available light filters in different colors also caught the attention (especially) of the two young children. The different colored filters and their transparency offered a multiplicity of layered possibilities, combined with the light table. The light filters and the two young children were responding and making decisions "in the *vibrant matter* that surround[ed] them and that [made] them" (Duhn & Quinones, 2018, p. 8). The large number of frames with multicolored filters might also have made a difference. The blackbox provided the children with tools that enabled them to study, explore, and listen more carefully to the materials' potentials, properties, propositions, materialities, and even objections.

When the two young children started combining the different materials on the light table, something unpredictable also happened (Figures 10.3 and 10.4). The different recycled materials became "more than they appeared to be at first glance" (Manning, 2013, p. 92). Until this moment, building on the floor and on the light table had been an individual activity, but in this building, two humans were involved, as well as different recycled materials with invitations and suggestions.

The environment in the blackbox, with its darkness and structured spaces for aesthetic explorations, also seemed to invite all the participants to slow down. The two young children were mostly quiet and appeared concentrated, only exchanging words about which materials they could add to the construction. Their building seemed attentive and in dialogue with the recycled materials and the matter of light, colors, and shadows. At one point, they were talking about building a city, but the story was more built than voiced, as the combination of the different materials "mobilized, provoked and shaped unfolding experimental activity" (Manning, 2013, p. 92). In these events (Figures 10.3 and 10.4), the objects were no longer just things but also forces with potentials, which could launch events that would challenge and open up to sense things and events with new eyes. The recycled materials, the light filters, the metal pieces, and other forms of materiality are not mere stuff. "Rather, matter is substance in its iterative intra-active becoming—not a thing, but a doing, a congealing of agency" (Barad, 2012, p. 80). The transformations from being objects to becoming objectiles make us concentrate more on what the materials *do* and less on what they *are*.

## New thoughts, questions, and imaginings through visual writing

My visual writings (Figures 10.1–10.4) emphasized how new experiences were produced by aesthetic provocations (Ulmer & Koro-Ljungberg, 2015, p. 142). In my encounters with images and videos from the study, words came to my mind that "communicated" with the materials' vibrancy and added yet another perspective when encountering the data; the visual writing with the events, the video sessions, the photos, and the readers of this chapter. I wrote about sensory and visual experiences in my encounters with the videos and the images, as well as how the different matter drew my attention. This may be a way of "a seeing through thought—thus simultaneously paradoxically seeing and not-seeing" (Ulmer & Koro-Ljungberg, 2015, p. 141). Such a paradox could be productive in this chapter because it offers ways that force me as a writer to encounter the invisible and imagine what I see (Ulmer & Koro-Ljungberg, 2015) through purposeful documentation. These encounters among young children, recycled materials, light tables, and materialities produced new thoughts, imaginings, and questions.

To think and to do in new materialism are challenging. The process embraces a flat ontology, where the researcher encourages a dual position and tries to reduce the performative self and open up to the different matter and materialities. Bennett (2010, p. ix) asks, "[I]s it not a human subject who, after all, is articulating this theory of vibrant matter?" This question is answered with both a *yes* and a *no*. Bennett continues, the "contradiction may well dissipate if one considers revision in operative notions of matter, life, self, self-interest, will and agency" (2010, p. ix). Although every agent may have its languages to narrate these experiences, I am the one who articulates them. This is not a perspective that I can set aside. These issues are also raised by Colebrook (2014) and Lenz Taguchi (2017), who picture a posthuman researcher as one who tries to write about other materialities and species in a flat, equal, and mutual kind of way, not centering his or her research around human thinking and doing. Nonetheless, there is considerable power in being the storyteller. Again, humans use a language that gives it meaning and importance through the narratives that only humans understand (Lenz Taguchi, 2017).

Inspired by Bennett's (2010) work, I mobilized a cultivated, patient, and sensory attentiveness to non-human forces. The entanglements of the dark room-spotlights-transparent-light-filters-metal objects-young children became especially interesting to explore. These narratives (visual writings) were written by myself yet together with the non-human forces and in rereading and writing in new encounters with the expressive images and videos. These different agents also contributed to our collective knowledge production.

## Imagining sustainable futures through encounters with recycled matter

By using the concepts of vibrant matter and objectiles, the previous sessions unfolded how the children were drawn to the recycled materials' vibrancy and seemed to act on the force from the objects. When the young children held these materials in their hands and built and aesthetically explored them, valuable connections between the young children and the materials were created. All encounters with materials set things in motion, evoke memories, narrate stories, invite actions, and communicate ideas (Pacini-Ketchabaw, Kind, & Kocher, 2017).

In this chapter, I also address the ethics about questioning the dualism between the human and the material subjects from the theoretical perspectives of vibrant matter and objectiles. According to Pacini-Ketchabaw et al. (2017, p. 1), "Every encounter demands its own questions, its own concerns, its own ethos." Bennett's (2010) vital materiality encourages humans' engagement with vibrant matter and draws attention to the potential ethical and political implications of such an engagement, particularly from an ecological perspective.

For example, in the creative reuse centers where I have worked, young children are allowed to encounter all materials. Hence, a young child could empty a whole bucket of buttons or make a lot of noise by beating a barrel with a metal stick. A teacher in such an environment plans for small groups and needs knowledge about the materials as active and participatory, as well as the children's multisensory ways of encountering them. Vecchi (Carlsen, 2015) encourages us not to perceive recycled materials in a simplified manner but based on an ecological mindset, where we emphasize the connections between the children and the materials and between the children and the environment. Furthermore, she emphasizes that these materials ought to be promoted by adults "who know the materials, techniques and the ideas' identities" (Interview in Carlsen, 2015, p. 46). Vecchi's thoughts bring me to my own experiences with recycled vibrant matter in our local creative reuse center:

> Standing by the kitchen sink in our creative reuse center. Hands in foam and water.
> Sink is filled with objects. Pieces of metal. Washing. Touching. Turning. Sensing.
> Smooth against my fingertips.
> Smell of metal and soap.
> Sparkling with the water.

The preceding short narrative unfolds the ethical and the valuable connections that Vecchi (Carlsen, 2015) discusses. A large part of my work as a project leader comprised washing, arranging, and exhibiting the materials for our visitors. At first, the washing sessions seemed endless and sometimes meaningless. After a while, it felt different. Through washing the objects, I came to know them—how they felt in my hands, smelled, and twinkled. The objects made me curious about their histories and earlier lives, as well as why they were thrown away. By arranging them, I understood their potentials, properties, and qualities; whether they could be stacked; and how they looked alone and in groups. By exhibiting them, I also learned about their relationships with other objects and how I felt that they looked inviting with some materials but not with others. By being acquainted with them, their identities and histories as industrial "waste" were told. All this knowledge was valuable when engaging in workshops with the children. All this knowledge could also be valuable for more sustainable futures. The recycled materials were not simplified and not regarded as worthless. On the contrary, they were presented in a creative, proactive, and optimistic way. Getting to know these materials could inspire humans to reconsider materialism (Parnell et al., 2017).

It is valuable to obtain first-hand knowledge of the endless amount of materials that were used only once or never used because they were damaged goods, out of season, or offcuts. These materials could include paper, metal, or wood offcuts from industries; seasonal goods, such as decorative objects; or different kinds of containers. By getting to know these materials, we could extend their lifetimes if we offered them to ECCs, schools, students, adults, and artists. Again, this knowledge could reduce consumption of new materials (such as toys, paper, building materials, and decorative items) in educational institutions, so we can contribute to and assume a common responsibility for our future. According to Duhn (2015), the more contacts among various vibrant matter are established, the greater the opportunities are for matter to form new entanglements and open up possible futures for thinking and doing. Hence, young children's aesthetic explorations with these materials could open up thoughts, talks, and imaginings.

On another level, ethics enable us to encounter and engage with recycled materials; question the western countries' consumption habits; and learn to value, take care of, inquire about, and be creative with discarded, imperfect, rejected, and presumed "worthless materials." Children's engagement, participation, and contribution "to the issues that affect their lives now and in the future" (Duhn, 2012, p. 21) could be ways of addressing and imagining sustainable futures. A visit to a Remida reveals the huge volume of materials and the consumption for which each town/municipality is responsible. On one hand, our common "trouble" and responsibility for our damaged Earth emphasize the need for change. On the other hand, the variety and the amount of recycled materials make a Remida an exciting place to visit. These complexities and paradoxes of our time may force us to think differently and be creative in new conceptual ways. Recycled materials force humans to acknowledge their responsibility through the items' visibility, volume, and variety, as "consumption and mass production is made visible" (Duhn, 2012, p. 26). Hence, working with multisensory thinking and doings with recycled materials' vibrancies could contribute to young children's knowledge about consumption and ability to participate in more sustainable futures on a community level. I have also argued earlier that working with these materials provides meaning in itself through a greater awareness of the ability to reuse things (Odegard, 2012).

## Final matters

In this chapter, I have challenged and discussed some aspects on which humanism rests because I find it unreasonable to use an approach based on the belief that humans' position in the world is superior to those of animals, things, and matter (Johansson, 2017, p. 136). Haraway's (2016) notion of *"staying with the trouble"* challenges the thinking about human sovereignty and the Anthropocene. When the human position is challenged, animals, things, matter, techniques, and humans exist on the same immanent level, with the same status and importance, possibilities, and preconditions (Johansson, 2017, p. 148). Recognizing human agency and responsibility in the Anthropocene and staying with the trouble connect with recycled materials. Based on my research for this chapter, working toward sustainable futures could involve encouraging children's new ideas and imaginative thinking, as well as rethinking about responsibility in relation to matter and about thought itself (Duhn, 2016, p. 383). These encounters offer possibilities for learning and change. Recycled materials could become matter that matters.

A few events with young children cannot make me evaluate and scientifically claim that children's work with recycled materials at present will make them more conscious consumers in the future. However, working with recycled materials for over a decade lends me some authority to claim that it could be realized—based on my own experiences and examples from young children's everyday lives that I have observed in ECCs. These stories tell me that young children choose to use recycled materials in their play and learning, and they think creatively about their use. They also use recycled materials to solve different challenges concerning the items they need in their everyday lives. It could also be meaningful, hopeful, and encouraging to think that it could make a difference, and this again could make us "think and imagine pedagogical practice differently" (Duhn, 2016, p. 379).

In this chapter, I have addressed encounters between humans and materials, especially between young children and recycled materials through the concepts of vibrant matter and objectiles. The events and my subsequent encounters with the images and the videos have fascinated and drawn me in, as well as made me think and ask the following questions: how can teachers provide environments that make these kinds of connections possible? How do we reposition ourselves and draw on children's multisensory ways of approaching materials? How do ECC teachers respond to young children's experimenting and exploring, as well as the possibilities with vibrant matter and objectiles? Bearing in mind my discussion and exploration in this chapter, a general openness to children's multisensory ways of responding to the vibrancy in the materials and the materials as objectiles could offer new possibilities. When young children empty a bucket of buttons or bang on a barrel, it could be just a playful, mischievous act. However, thinking about this act with new materialistic ideas could expand it to a young child's response to the button's call and the barrel's vibrancy by experimenting with, exploring, and encountering the materials and their materialities as objectiles.

Anzaldúa's (1987) quote in the beginning of this chapter asks us when we started to objectify things and lose touch with them. Through a study of young children's aesthetic explorations, experimenting, and narrating, I have attempted to explore how humans can reconnect with objects by studying human lives as embedded in a material world of high complexity (Hayles, 1999, p. 5). From this endeavor, they can build knowledge on how to live together in the more-than-human worlds (Haraway, 2008).

## Note

1 An atelierista is an early childhood teacher/teacher who often has a background in the arts. The atelierista in a Reggio Emilia Early Childhood Center is a studio teacher in an atelier with plenty of open-ended and explorative materials and tools, such as paint, clay, and recycled or natural

materials. The atelierista performs a complex role with many different tasks, including supporting children and teachers in their encounters with materials and the many languages of expression. The atelierista is responsible for creating or following up on different projects, introducing new concepts to the children, and sparking their curiosity, creativity, and aesthetic explorations.

## Acknowledgments

I am grateful for all the recycled materials that have made a difference in my life. I appreciate the young children, the artist, the atelierista, and the early childhood teachers for participating and sharing their experiences with me in this research. I also thank the editors, the reviewers, and my supervisors for their thoughtful, inspiring, and enthusiastic comments along the way.

## References

Alaimo, S. (2018). Material feminism in the Anthropocene. In C. Åsberg & R. Braidotti (Eds.), *A feminist companion to the posthumanities* (pp. 45–54). Cham: Springer International Publishing.

Anzaldúa, G. E. (1987). *Borderlands, La Frontera: the new Mestiza* (3rd ed.). San Francisco, CA: Aunt Lute.

Barad, K. (2012). Intra-actions. An interview with Karen Barad by Adam Kleinman. *Mousse, 34*, 76–81.

Barad, K. (2008). Posthumanist performativity: Toward an understanding of how matter comes to matter. In S. Alaimo & S.J. Hekman (Eds.), *Material feminisms* (pp. 120–157). Bloomington, IN: Indiana University Press.

Bennett, J. (2010). *Vibrant matter: A political ecology of things*. Durham, NC and London: Duke University Press.

Carlsen, K. (2013). "Å skape forbindelser." Atelierets plass og atelieristens kompetanse i barnehagene i Reggio Emilia. *Techne Series A, 20*(1), 13–39.

Carlsen, K. (2015). *Forming i barnehagen i lys av Reggio Emilias atelierkultur*. Åbo: Åbo Akademi.

Ceder, S. (2016). *Cutting through water: Towards a posthuman theory of educational relationality*. Lund, Sweden: Lund University.

Colebrook, C. (2012). *Death of the posthuman: Essays of extinction. Vol 1*. Ann Arbor, MI: Open Humanity Press.

Colebrook, C. (2014). *Sex after life: Essays of extinction. Vol 2*. Ann Arbor, MI: Open Humanity Press.

Crutzen, P. J. (2006). The "Anthropocene". In E. Ehlers & T. Krafft (Eds.), *Earth system science in the anthropocene: Emerging issues and problems* (pp. 13–18). Berlin, Heidelberg: Springer.

Deleuze, G. (1993). *The fold: Leibniz and the Baroque* (T. Conley, Trans. Vol. 1). Minneapolis, MN: University of Minnesota Press.

Duhn, I. (2012). Making "place" for ecological sustainability in early childhood education. *Environmental Education Research, 18*(1), 19–29.

Duhn, I. (2015). Making agency matter: Rethinking infant and toddler agency in educational discourse. *Discourse: Studies in the Cultural Politics of Education, 36*(6), 920–931.

Duhn, I. (2016). Speculating on childhood and time, with Michael Ende's Momo (1973). *Contemporary Issues in Early Childhood, 17*(4), 377–386.

Duhn, I., & Quinones, G. (2018). Eye-to-eye with otherness: A childhoodnature figuration. In A. Cutter-Mackenzie, K. Malone, & E. Barratt Hacking (Eds.), *Research handbook on childhoodnature* (pp. 1–16). Cham: Springer International Handbooks of Education.

Haraway, D. J. (2008). *When species meet* (Vol. 3). Minneapolis, MN: University of Minnesota Press.

Haraway, D. J. (2016). *Staying with the trouble: Making kin in the Chthulucene*. Durham, NC & London: Duke University Press.

Hayles, N. K. (1999). *How we became posthuman: Virtual bodies in cybernetics, literature, and informatics*. Chicago, IL: University of Chicago Press.

Hird, M. J. (2012). Knowing waste: Towards an inhuman epistemology. *Social Epistemology, 26*(3–4), 453–469.

Holloway-Attaway, L. (2018). Embodying the posthuman subject: Digital humanities and permeable material practice. In C. Åsberg & R. Braidotti (Eds.), *A feminist companion to the posthumanities* (pp. 91–101). Cham: Springer International Publishing.

Johansson, L. (2017). Rörelsens pedagogikk: Att tillvarata lärandets kraft [Movement's pedagogy: To utilize the power of learning]. In B. Bergstedt (Ed.), *Posthumanistisk pedagogikk. Teori, praksis, og forksningspraktik [Post-humanist pedagogy. Theory, practice, and research practice]* (pp. 133–149). Malmö: Gleerups utbildning AS.

Kind, S. (2014). Material encounters. *International Journal of Child, Youth and Family Studies, 5*(4.2), 865–877.

Lafton, T. (2015). Digital literacy practices and pedagogical moments: Human and non-human intertwining in early childhood education. *Contemporary Issues in Early Childhood, 16*(2), 142–152.

Lenz Taguchi, H. (2017). Ultraljudsforsterbilden: Et feministisk omkonfigurering av begreppet posthumanism. In B. Bergstedt (Ed.), *Posthumanistisk pedagogik. Teori, praksis og forskningspraktik [Post-humanist pedagogy. Theory, practice, and research practice]* (pp. 167–190). Malmö: Gleerups utbildning AS.

Manning, E. (2013). *Always more than one: Individuation's dance.* Durham, NC: Duke University Press.

Manning, E. (2016). Weather patterns, or how minor gestures entertain the environment. In U. Ekman, J. D. Bolter, L. Diaz, M. Sondergaard, & M. Engberg (Eds.), *Ubiquitous computing, complexity and culture* (pp. 48–57). New York & London: Routledge.

Matter. (2018). Retrieved from www.merriam-webster.com/dictionary/matter. Retrieved 10.10.2018.

Odegard, N. (2012). When matter comes to matter: Working pedagogically with junk materials. *Education Inquiry, 3*(3), 387–400.

Odegard, N. (2015). *Gjenbruk som kreativ kraft. Når (materi)AL(ite)T henger sammen med alt. [Reuse as a creative force. When matter comes to matter].* Oslo: Pedagogisk forum.

Odegard, N. (2019). Crows: Young children's aesthetic explorations of crow. In Pauliina Rautio & Elina Stenvall (Eds.), *Social, material and political constructs of Arctic childhoods: An everyday life perspective* (pp. 119–137). Singapore: Springer Singapore.

Odegard, N., & Rossholt, N. (2016). In-betweens spaces. In A. B. Reinertsen (Ed.), *Becoming earth: A post human turn in educational discourse collapsing nature/culture divides* (pp. 53–63). Rotterdam: SensePublishers.

Pacini-Ketchabaw, V., Kind, S., & Kocher, L. L. M. (2017). *Encounters with materials in early childhood education.* New York: Routledge.

Parnell, W., Downs, C., & Cullen, J. (2017). Fostering intelligent moderation in the next generation: Insights from Remida-inspired reuse materials education. *The New Educator, 13*(3), 234–250.

Pink, S. (2007). *Doing visual ethnography: Images, media and representation in research.* London: Sage.

Pink, S. (2015). *Doing sensory ethnography* (2nd ed.). Los Angeles, CA, London, New Dehli, Singapore, Washington DC: Sage.

Remida. (2018). Reggio children foundation. Retrieved from http://remida.reggiochildrenfoundation.org/?lang=en.

Somerville, M. (2013). *Water in a dry land: Place-learning through art and story.* London and New York: Routledge.

Somerville, M. (2015). Children, place and sustainability. In M. Somerville & M. Green (Eds.), *Children, place and sustainability* (pp. 166–184). London: Palgrave Macmillan.

Taylor, C. A. (2016). Edu-crafting a cacophonous ecology: Posthumanist research practices for education. In C. Hughes & C.A. Taylor (Eds.), *Posthuman research practices in education.* London: Palgrave Macmillan.

Tesar, M., & Arndt, S. (2016). Vibrancy of childhood things. Power, philosophy and political ecology of matter. *Cultural Studies Critical Methodologies, 16*(2), 193–200.

Ulmer, J. B., & Koro-Ljungberg, M. (2015). Writing visually through (methodological) events and cartography. *Qualitative Inquiry, 21*(2), 138–152.

Vecchi, V. (2010). *Art and creativity in Reggio Emilia: Exploring the role and potential of ateliers in early childhood education* (Vol. 8). Abingdon: Routledge.

Vecchi, V. (2012). *Blå blomster, bitre blader* (L.H. Støyva, Trans.). Bergen: Fagbokforlaget Vigmostad & Bjørke AS.

# 11

# GARDENING WITH CHILDREN AND PRE-SERVICE TEACHERS

## Considering terrestrial collective(s) in action

*Janice Kroeger, Terri Cardy, Abigail E. Recker, Lynn Gregor, Aubrey Ryan, Anna Beckwith and Jacob Dunwiddie*

## Considering *sustainable futures* with young children

*Sustainable futures* is a term used by some early childhood scholars throughout the world as each considers what it might mean for young children and others to care for the world and other worldly creatures in the early years and education of young children (Bone, 2010; Galvez et al., 2018; Hedefalk, Almqvist, & Östman, 2014). About such work, Bruno Latour comments (2018), "Humans are no longer the only actors, even though they still consider themselves entrusted with a role that is much too important for them" (p. 43). In the collective action/social action described herein, a small group of individuals from undergraduate and graduate education, lab school faculty, and international and local elders interested in gardening in schools worked together, sharing an ideological passion and desire for children to work *with* the Earth as a form of social and political caring (Lash & Kroeger, 2018).

Particularly, our group was enacting the concept of sustainable futures (SF) as others have articulated this idea in various ways, caring for not just each other but the Earth and worldly others (Duhn, 2012; Ritchie, 2016). For example, Outdoor Educator, Terri Cardy has for several years and counting led a zero-waste policy initiative at our Child Development Center, transforming lunches and snacks, omitting single use plastics and enacting composting with all food waste for 126 children. The Center also enacts the use of *Remida* in children's constructions (see Odegard this edition) as many other laboratory schools in North America do (Parnell, Dowens, & Cullen, 2017); Cardy's work with the entire staff at the Center fosters a focus on outdoor education on playground, within gardening and in an adjacent meadow and wetlands (Galizio, Stoll, & Hutchins, 2009; Sisson & Lash, 2017; Porto, this edition). Additionally, Environmental Educator Lynn Gregor was working on gardening projects in several local elementary schools in our surrounding communities (Hassler, Gregor, & Snyder, 1999).

Simultaneous to Cardy and Gregor joining our project, Kroeger, with several of her students, worked with permaculture educator and elder Solomon Amuzu (Raimbekova & Amuzu, this edition). We were building alliances across the globe through social media, geared toward motivating new teachers to consider the Earth in their daily curriculum endeavors in public schools. The convergence of our sustainability work is described here as a sustained activity, focused not just on the agency of humans (Lash & Kroeger, 2018), but as a wider set of social ties and "definition of associations that make up" our more than human "collectives" (Latour, 2018, p. 57; Taylor, 2018; Taylor, Blaise, & Giugni, 2012; Taylor & Giugni, 2012).

While we in northeast Ohio, did not enact sustainable futures pedagogy in the same ways as we saw our international colleagues doing so, we crafted our own work based upon social action models described elsewhere (Kroeger & Lash, 2005). We became more fully aware of other early childhood centers as they focus on an understanding of worldly creatures within the milieu of children, and our changed actions and mentality began surfacing in similar time frames as others across the globe (Bone, 2010; Pacini-Ketchabaw & Nxumalo, 2018; Pacini-Ketchabaw et al., 2015; Taylor, 2018; Taylor & Giugni, 2012). This desire for *worlding* (Haraway, 2016; Latour, 2018) converged with the work of students in program(s) and our colleagues Cardy and Gregor as we merged our desires to promote sustainable future(s) within the task of early childhood and teacher education (Taylor & Giugni, 2012; Taylor et al., 2012).

## Social action: Including the Earth with young children's learning

Initially, Kroeger invited undergraduates to join her in social action planning, and a small group converged to discover and plan how earthly matters could figure into kindergarten through third grade curriculum. We began by promoting curriculum planning involving gardening and community action with then undergraduate students and a small group of students, Aubrey Ryan, Jacob Dunwiddie, and Kaylee Klink. Pre-service teachers volunteered to meet with Kroeger for nearly 18 months, in various capacities, to include the Earth in curriculum; such work involved monthly meetings and goal setting—the work preservice teachers did was beyond what was expected and not associated with a grade for a class or even with Kroeger as their professor, but merely as a faculty member in their program. During a fall field experience, Dunwiddie and Ryan chose to work in their local field placements to make social connections by helping young children to communicate across the globe and learn via Solomon Amuzu with distance technology (email, Facebook, digital photo exchanges). Several teachers developed curriculum with gardening: Ryan with literacy, communication, writing and reading, mathematics, and graphing; Dunwiddie with mathematics, measuring area, and indoor germination projects (Figure 11.1); and Kaylee Klink with milkweed germination of the Monarch Butterfly habitat.

We recognize here that many of the assumptions within teacher education are instrumentalist—focused on testing, academic standards, and performance goals, especially in North America. However, the conversations between this small group of preservice teachers and professor were more than that, even if the curriculum work they performed appeared rather conventional. Dunwiddie for example strove to integrate environmental education elements in much of his teaching, despite the lesser standards in his evaluations and for state licensing; Ryan continued to supplement traditional teaching with social justice and eco-pedagogy imperatives.

Integrating sustainable futures in a field experience with refugee and immigrant children and families, Aubrey Ryan reached out via social media to Solomon Amuzu, providing children with a Ghanaian community elder who expertly demonstrated the connection between the Earth, production gardens for children in another part of the world, and the importance of relations between soil, irrigation systems, water catchment, chickens (their wastes), and local trees. Children's work in the North American classroom produced

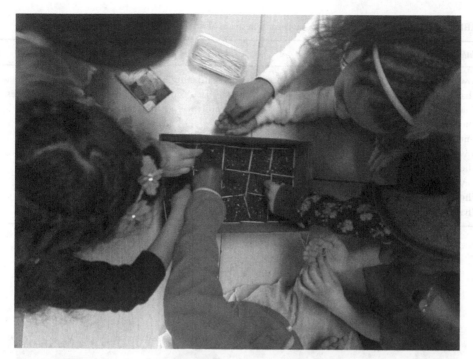

**FIGURE 11.1** Dunwiddie's work with children on germination and calculating area
*Source*: Jacob Dunwiddie

associations of meaning between food and clean climate, as well as a deeper sense of community in the urban classroom as Ryan and young children developed a relationship with Solomon and the children in Ghana. By working together across several weeks of curriculum development, children who were in urban schools (many coming from warmer climates like Solomon's) engaged with tasting fruits and vegetables, planting their own seedlings for home gardens, developing question forms and finding commonalities within their communities and his. The first graders composed a group letter and Ryan supported their pen pal exchange with Solomon over a month of planning. Children's engagement with meaningful curriculum urban environment was noteworthy, as each learned writing and reading, but also as some identified with climates across the globe and because the spaces and histories around foods the children shared exposed deeper geographic and social connections (see Kenyon, Coffee, & Kroeger, 2016; Kroeger, Amuzu, & Ryan, 2017). Solomon answered a group letter via email and shared images via social media, even as questions about his farming practices and local children's desires for fruits and vegetables from their home countries surfaced (see Table 11.1).

Through six months of working together we met, developed goals, planned, and strategically built reciprocity with Amuzu (Figure 11.2). As undergraduate Ryan, Dunwiddie, and Klink completed field assignments in curriculum around spring planting, discovering the life-cycle, and exploring with young children the relations between food, clean water, and food production, we were also planning together how we could ultimately help Amuzu continue his own agenda of building sustainable futures in Ghana (Raimbekova & Amuzu, this edition). Our local actions were fundraising for Amuzu (eventually securing a laptop computer), distribution of local seeds, and procuring classroom material; particularly, Kaylee Klink created an early milkweed planting project for pollinators and an endangered

**TABLE 11.1** Ryan's project: children's written questions about food production gardens in Ghana

| Children's questions | Children's reflections |
| --- | --- |
| How long does it take to grow your crops? | I like gardening because you can make the world a |
| What is your favorite crop to grow? | better place and because I like fruits and vegetables. |
| What do you like about gardening? | My family has a garden, my job is to water the |
| Where do you get the water to water your | plants. |
| plants? | I love your garden so much! |
| How do you water your plants? | I love to make the Earth a better place. |
| Where do you get your seeds? | I think gardening is amazing. |
| Do you get tired after gardening? | I like your garden because it is very big and very |
| Who are your helpers? | special! |
| What are some of your favorite foods to eat? | Hello, I am from Congo in Africa, I love Africa! |

**FIGURE 11.2**  Amuzu with farmers' market banner

*Source*: Janice Kroeger

species of insect in our region, the Monarch butterfly. We found a way to transport a new laptop to Amuzu via a group of gardeners travelling to Ghana from the University of Wisconsin Madison, and in late May of the year we started, we planned the next phase of our evolving social action project. The group planned how and what we would do if Amuzu could join us in North America the following Spring when Dunwiddie, Ryan, and Klink were in their final student teaching. For example, we created a plan to use our local farmers' market as a space to share our political work supporting Amuzu and children in Ghana, while also teaching children in North America about understanding the Earth.

Modest profits from local markets selling Amuzu's Moringa Mint Tea allowed us to fund aspects of Amuzu's travels, get the word out about our work in local schools, and involve teacher education students in local community action tied to the values of sustainability and alternative forms of social exchange.

As Kroeger searched for funding and school sites for expanded interest of other local master gardeners and naturalists we also began planning with our two local experts, Terri Cardy and Lynn Gregor, and then graduate students Abbey Recker and Anna Beckwith. With a larger group joining our Social Action planning and sustainability efforts, we gained the ear and support of figures in our local school district and university. The linkages of more expertise among our working group allowed our efforts of sustainability to grow. For example, Recker and Beckwith shadowed Lynn Gregor as she worked at Holden Elementary school with primary school children in kindergarten to 5th Grade in their gardens, in the classroom, and within the community. Recker spearheaded the development of a gardening curriculum for wider distribution for educators in North America in concert with the work of science educator, Dr. Bridget Mulvey with Kroeger. Anna Beckwith collected and annotated a children's literature collection as teaching materials to complement Recker's science, while also learning from and assisting Environmental Educator, Lynn Gregor. Anna's materials supplemented our gardening efforts and built our local sustainability resources. When Amuzu came to North America to visit our school gardens, curriculum materials for prekindergarten to grade 5 materials were completely available to in-service as well as pre-service teachers via website. And, with additional funding finally secured through the Gerald Read International Center and the College of Education, Health and Human Services and the University Teaching Council we planned a teaching symposium

**FIGURE 11.3** Garbage from cleaning meadow
*Source*: Janice Kroeger

for in-service and pre-service teachers, as well as a series of school visits across an entire week. Our events converged right around Earth Day in North America, when Amuzu joined us in person.

The events we planned for Amuzu's visit were geared toward teachers and other educators wishing to promote gardening in schools. Additionally, Amuzu's visit included several days in our Center, working with very young children (cleaning up the meadows, singing and dancing, sharing Ghanian songs and Earth stories, playing musical instruments from Ghana, turning the soil, planting, and developing compost (see Figures 11.3, 11.4, 11.5).

Gregor invited Amuzu, Recker, and Kroeger to hear the efforts of grade school children, as experts in their own knowledge construction, conserving water, growing and selling food while learning, and having local impact on community farmers' markets. Amuzu also worked with many groups of older children at Helden Elementary School in the form of international

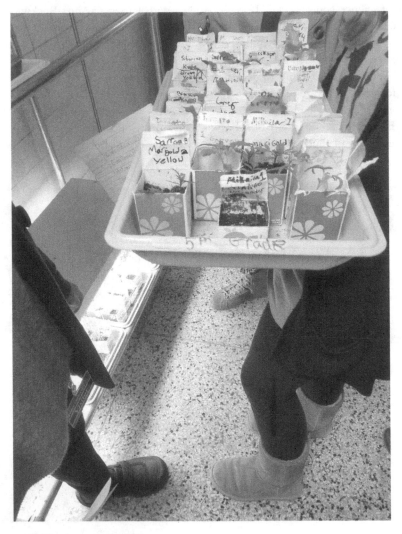

**FIGURE 11.4** Helden images gardening
*Source*: Janice Kroeger

**FIGURE 11.5**   Turning the compost
*Source*: Janice Kroeger

guest speaker. Older children shared their school garden projects under the leadership of environmental educator, Lynn Gregor. Amuzu was given a school tour, as students shared their work with small grow beds, heat lamps for germination and indoor planting, and water catchment systems for outdoor work.

Amuzu, in exchange, answered questions about the problems of sustainability in Ghana, gardening in a hot climate, supporting water quality in his home country, a country with a colonialized past and one experiencing current illegal mining practices. Elementary classrooms, one age group at a time, heard the message of the importance of caring for the planet; particular teachers in the school supplemented their weekly work by integrating geography and science into materials. Students at Helden Elementary also asked Amuzu questions about sustainability and conservation in Ghana, mirroring their understanding of that concept in North America.

Our week of Solomon Amuzu's visit culminated with a larger series of events for in-service and pre-service teachers and built Amuzu's networks in North America. During our symposium for teachers, for example, panel members Cardy, Gregor, and Amuzu spoke about the importance of integrating environmental education and sustainability into curriculum. Gardening as one venue for that mission was shared.

Our now growing network gave away supportive curriculum materials, collected questions about environmental education for blog postings, described tips for gardening in schools, and developed a sophisticated informational session, bringing our Social Action Project to a modest but successful culmination. While in North America, Amuzu was able to visit an important regional environmental education center, a national park, meet with several future business partners, and finally visit a local Green Partnership in an Urban

Agriculture and Youth Education Program. Elsewhere we've argued that social action builds associations in unpredictable ways. After his visit, Amuzu began consulting with an heirloom seed company, building his own global connections, allowing him to revisit the schools we started with and following up with children's questions about the Earth. As part of an ongoing project merging technology, communication, and master gardeners, Amuzu continues to support Helden in their gardening work and we are planning another visit from Amuzu in the near future.

## Priorities in gardening with Outdoor Educator, Terri Cardy

In the following section of this chapter, Outdoor Educator Terri Cardy shares considerations for working with young children outdoors, especially within gardening. Gardening is considered one of the most elemental aspects of outdoor education for young children and is widely practiced around the globe. This section of the chapter is transcribed from her talk. Her work was an important part of our panel of speakers during Amuzu's Earth Week visits, and Cardy's words are illustrated with photos from Amuzu's visit with us in North America.

### Respecting all things

When children participate in gardening they experience first-hand the cycles of life that occur throughout the year. Gardens are a space for children to experience the wonder of new life and explore possibilities in learning. Exploring can be in a large vegetable garden, an orchard of trees, or a small herb garden to taste, touch, and enjoy the smells. Gardens help children discover life. And not only do they gain an appreciation of where food comes from, but they also get to be caretakers of something smaller than themselves. Children can learn that with love and attention beautiful things can grow (Keeler, 2008). Gardening affords children the opportunity to share their thoughts and ideas with those around them, to make meaning of the world around them. Children learn respect for all living things in outdoors spaces like gardens. Many children want to take care of living creatures. Those children who haven't yet developed care and empathy for other living creatures can see it modeled by teachers and other children. With encouragement and guidance, children learn to take care of living plants as well. With the abundance of living things, children have almost unlimited opportunities to engage in respectful and caring actions towards many living things. While working in the garden with a group of ten preschoolers one fall morning, a four-year-old looked at me and said, "Mrs. Cardy we have to build a home for the worm. It will stay safe then." This child was concerned that the worm would become injured as the children using metal trowels began digging up onions around the worm.

### Doing real work in the world

The idea of "real work" always comes to surface when working with young children. Even children as young as 18 months show adults just how capable they are as they use real tools to do real work. A garden area is the perfect setting for children to pull weeds, carry heavy rocks, push filled wheelbarrows, and water plants, as well as other tasks. This real work is another way that children exert power with the garden, and the garden works back. This real work is a means for developing large motor skills and for children to learn their capabilities, as well as to understand the power of the things with which they interact.

**FIGURE 11.6** Planting with young children
*Source*: Janice Kroeger

Children experience a sense of joy and achievement in working collaboratively to move a large pile of topsoil, to spread a bale of straw, or to sow seeds (see Figure 11.6). As teachers, we must trust that children are capable and serious workers.

If we hope for children to become fully engaged learners, we as adults must not plan *for* the children, but instead *with* the children and with the materials themselves. Through careful decision making with the children the garden can be a place where rights are shared. The children will develop a relationship with this space, with these creatures and things, and in turn valuing and collaboration occurs. With this sense of place, children may want to share accomplishments with others including classmates, family, and the larger community. Your garden may be as simple as a single apple tree, which has many very complex elements to a young child. Or your garden may have several and varied plantings entwined with pathways, water features, and pieces of artwork. Really working within the garden with children can be as simple or as elaborate as you and the children decide, while keeping in mind your space, skills, and materials and larger objectives. Even the smallest garden is a learning opportunity for the youngest children. When gardening with young children it isn't about large production, but instead giving children the opportunity to understand how the world works. Think about what a two-year-old can learn from growing a tomato plant from seed to the plant's duration. What will a three-year-old learn from planting a milkweed plant and later finding a Monarch butterfly egg on it as she delights in observing the transformation from caterpillar to chrysalis then to butterfly! Imagine a group of children finding that there are too many cucumbers, zucchini, or squash to eat, and when you asked them what we can do with all of these vegetables before they rot, to your delight, the children decide to give their produce to a farmer for her chickens or to a local food pantry for people without enough food?

## Earthly collectives change our conceptualization of terrestrial politics in social action

In a social action project, as conceptualized prior, noteworthy features of change happen best as the convergence of like-minded individuals of very different statuses work toward a common, heretofore, mostly humanist goal (Kroeger & Lash, 2005; Lash & Kroeger, 2018). What is new to social action here is the *terrestrial*, and yet an evolving idea that the Earth and materials themselves are actors with agency equal to or more monstrous (and/or perhaps sometimes more generous) than humans deserve (Latour, 2018). Adults were surprised for example, at how much waste was collected in the meadow adjacent to the Center and the tenacity with which very young children engaged in what they perceived as cleaning up for the ducks in the wetlands. Adults were surprised at the sophistication of Helden elementary student's encounters with nature as demonstrated by such questions to Amuzu as: *"What do you think of GMOs?"*

Because, our earlier considerations of social action were narrow, *worlding*, or world action seems now to be a more generative concept than social action (Lash & Kroeger, 2018). Haraway's arguments suggest that the concerns of planet and climate, human and otherwise, intersect as *entanglements*. In turn, thinking of humans in a *post*-world makes teacher education a shared future with the environment, for better or for worse. Latour argues that by thinking of *Terrestrial* (italics ours), we are "bound to the earth and to land" … aligned with "no borders," and transcending "all identities" (p. 54). We recognize now that collectives like the one shared in this chapter will change early childhood *qualities* and our politics as teachers; we must become Earth advocates and terrestrial activists instead of simply early childhood advocates (Latour, 2018; Ritchie, 2016). We turn again to Affrica Taylor's (2018) work, which has called for trying in our modest ways to (lose) the conceit that all action "is not just about us" (p. 206).

## References

Bone, J. (2010). Play and metamorphosis: Spirituality in early childhood settings. *Contemporary Issues in Early Childhood, 12*(4), 417–420.

Duhn, I. (2012). Making "place" for ecological sustainability in early childhood education. *Environmental Education Research, 18*(1), 19–29.

Galizio, C., Stoll, J., & Hutchins, P. (2009). "We need a way to get to the other side!" Exploring the possibilities for learning in natural spaces. *Young Children, 64*(4), 42–48.

Galvez, S. Duhn, I., Grieshaber, S., Odegard, N., & Penfold, L. (2018). *(Un)doing stories of quality in early childhood with teacups, turtles, and plastic.* 26th Annual Reconceptualising Early Childhood Education. Copenhagen, Denmark.

Haraway, D. J. (2016). *Staying with the trouble: Making kin in the Chthulucene.* Durham, NC: Duke University Press.

Hassler, D., Gregor, L., & Snyder, D. (1999). *A place to grow: Voices and images of urban gardeners.* Pasadena, TX: Pilgrim Publishing.

Hedefalk, M., Almqvist, J., & Östman, L. (2014). Education for sustainable development in early childhood education: A review of the research literature. *Environmental Education Research, 21*(7), 975–990.

Keeler, R. (2008). *Natural playscapes: Creating outdoor play environments of the soul.* Redmond, WA: Exchange Press.

Kenyon, E., Coffee, C., & Kroeger, J. (2016). "Hey, I've been there!" Using the familiar to teach world geography in a kindergarten classroom. *Social Studies and the Young Learner, 29*(2), 4–7.

Kroeger, J., Amuzu, S., & Ryan, A. (2017). *Gardening with Ghana.* Local to Global Justice Forum & Festival. Tempe, AZ: Arizona State University.

Kroeger, J., & Lash, M. (2005, October). Academics as/of diaspora: Changing our educational discourses. Social, action, projects: What is the justice of the social action project? [Daily plenary session]. *The 13th Conference: Reconceptualizing early childhood research, theory, and practice. "Language(s) in childhood(s)"*, University of Wisconsin, Madison, WI.

Lash, M., & Kroeger, J. (2018). Seeking justice through social action projects: Preparing teachers to be social actors in local and global problems. *Policy Futures in Education, 16*(6), 691–708.

Latour, B. (2018). *Down to earth: Politics in the new climate regime.* Cambridge, UK: Polity Press.

Pacini-Ketchabaw, V., & Nxumalo, F. (2018). Posthumanist imaginaries for decolonizing early childhood praxis. In M. Bloch, G. Swadener, & B. Canella (Eds.), *Reconceptualizing early childhood education—a reader: Critical questions, new imaginaries & social activism*, pp. 215–226. New York: Peter Lang.

Pacini-Ketchabaw, V. Nxumalo, F., Kocher, L., Elliot, E., & Sanchez, A. (2015). *Journeys: Reconceptualizing Early Childhood Practices through Pedagogical Narration.* Toronto: University of Toronto Press.

Parnell, W., Dowens, C., & Cullen, J. (2017). Fostering intelligent moderation in the next generation: Insights from Remida-inspired reuse materials education. *The New Educator, 13*(3), 234–250.

Porto, A. (2019). Nature can be dead and alive: Pachysandra is a bad guy. In J. Kroeger & C. Myers (Eds.) *Nurturing nature and the environment with young children: Children, elders, earth*, pp. 107–143. London and New York: Routledge (this edition).

Raimbekova, L., & Amuzu, S. (2019). The Call to Nature Permaculture project. In J. Kroeger & Casey Myers (Eds.) *Nurturing nature and the environment with young children: Children, elders, earth*, pp. 73–79. London and New York: Routledge (this edition).

Ritchie, J. (2016). Qualities for early childhood care and education in an age of increasing superdiversity and decreasing biodiversity. *Contemporary Issues in Early Childhood, 17*(1), 78–91.

Sisson, J. H., & Lash, M. (2017). Outdoor learning experiences connecting children to nature. *Young Children, 72*(4), 8–16.

Taylor, A. (2018). Situated and entangled childhoods: Imagining and materializing children's common world relations. In M. Bloch, G. Swadener, & B. Canella (Eds.), *Reconceptualizing early childhood education—a reader: Critical questions, new imaginaries & social activism*, pp. 205–214. New York: Peter Lang.

Taylor, A., Blaise, M., & Giugni, M. (2012). Haraway's "bag lady story-telling": Relocating childhood and learning within a "post-human landscape." *Discourses: Studies in the Cultural Politics of Education, 34*(1), 48–62.

Taylor, A., & Giugni, M. (2012). Common worlds: Reconceptualising inclusion in early childhood communities. *Contemporary Issues in Early Childhood Education, 13*(2), 108–120.

# INDEX

Printed in the United States
by Baker & Taylor Publisher Services